Chapter 3 Energetics, rates and equilibrium

Chapter 4 Organic chemistry, analysis and the environment

Specification list

AQA AS Chemistry

MODULE	SPECIFICATION TOPIC	CHAPTER REFERENCE	STUDIED IN CLASS	REVISED	PRACTICE QUESTIONS
AS Unit 1 (M1) Foundation chemistry	*Atomic structure*	*1.1, 1.2, 1.3, 1.4*			
	Amount of substance	*1.4, 1.5, 1.6*			
	Bonding	*2.1, 2.2, 2.3, 2.4, 2.5*			
	Periodicity	*2.6*			
	Introduction to organic chemistry	*4.1*			
	Alkanes	*4.2, 4.3*			
AS Unit 2 (M2) Chemistry in action	*Energetics*	*3.1, 3.2, 3.3*			
	Kinetics	*3.4, 3.5*			
	Equilibria	*3.6, 4.8*			
	Redox reactions	*1.7*			
	Group 7, the Halogens	*2.8*			
	Group 2, the alkaline earth metals	*2.7*			
	Extraction of metals	*2.9*			
	Haloalkanes	*4.3, 4.6, 4.8*			
	Alkenes	*4.4*			
	Alcohols	*4.5, 4.8*			
	Analytical techniques	*4.7*			

Examination analysis

AS Chemistry comprises two unit tests. All questions are compulsory.
Practical and investigative skills will also be assessed.

Unit 1	*Structured questions: short and extended answers*	*1hr 15min test*	*33 1/3%*
Unit 2	*Structured questions: short and extended answers*	*1hr 45min test*	*46 2/3%*
Unit 3	*Internal assessment of practical and investigative skills*		*20%*

The AS/A2 Level Chemistry course

AS and A2

All Chemistry A Level courses being studied from September 2000 are in two parts, with three separate modules in each part. Students first study the AS (Advanced Subsidiary) course. Some will then go on to study the second part of the A Level course, called A2. Advanced Subsidiary is assessed at the standard expected halfway through an A Level course: i.e., between GCSE and Advanced GCE. This means that new AS and A2 courses are designed so that difficulty steadily increases:

- AS Chemistry builds from GCSE science
- A2 Chemistry builds from AS Chemistry.

How will you be tested?

Assessment units

For AS Chemistry, you will be tested by three assessment units. For the full A Level in Chemistry, you will take a further three units. AS Chemistry forms 50% of the assessment weighting for the full A Level.

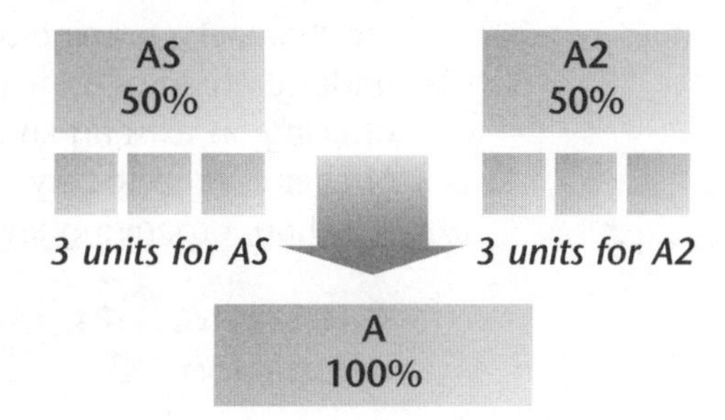

One of the units in AS and in A2 is practically based. You will take two theory units in each of AS and A2. Each unit can normally be taken in either January or June. Alternatively, you can study the whole course before taking any of the unit tests. There is a lot of flexibility about when exams can be taken and the diagram below shows just some of the ways that the assessment units may be taken for AS and A Level Chemistry.

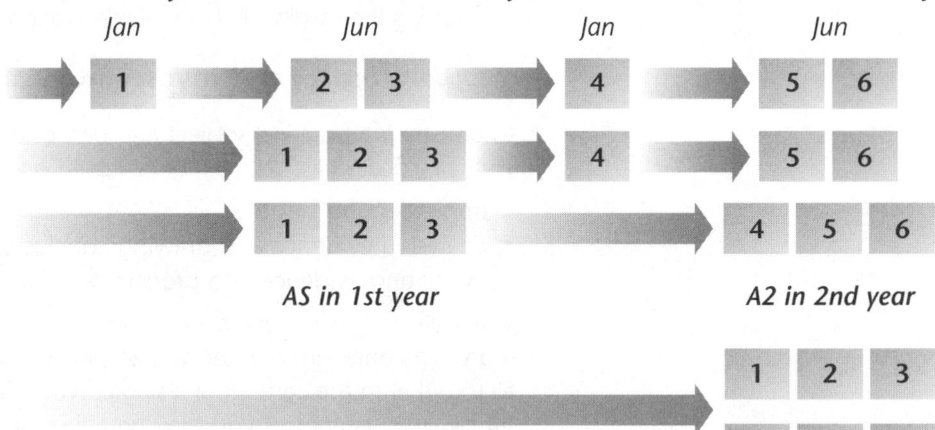

If you are disappointed with a module result, you can resit each module once. You will need to be very careful about when you take up a resit opportunity because you will have only one chance to improve your mark. The higher mark counts.

A2 and Synoptic assessment

After having studied AS Chemistry, you may wish to continue studying Chemistry to A Level. For this you will need to take three further units of Chemistry at A2. Similar assessment arrangements apply except some units, those that draw together different parts of the course in a synoptic assessment, have to be assessed at the end of the course.

Coursework

Coursework may form part of your A Level Chemistry course, depending on which specification you study. Where students have to undertake coursework, it is usually for the assessment of practical skills but this is not always the case. See pages 4–9.

Key Skills

To gain the key skills qualification, which is equivalent to an AS Level, you will need to collect evidence together in a 'portfolio' to show that you have attained a satisfactory level in Communication, Application of number and Information technology. You will also need to take a formal testing in each key skill. You will have many opportunities during AS Chemistry to develop your key skills.

What skills will I need?

The assessment objectives for AS Chemistry are shown below.

Knowledge with understanding

- recall of facts, terminology and relationships
- understanding of principles and concepts
- drawing on existing knowledge to show understanding of the responsible use of chemistry in society
- selecting, organising and presenting information clearly and logically

Application of knowledge and understanding, analysis and evaluation

- explaining and interpreting principles and concepts
- interpreting and translating, from one form into another, data presented as continuous prose or in tables, diagrams and graphs
- carrying out relevant calculations
- applying knowledge and understanding to familiar and unfamiliar situations
- assessing the validity of chemical information, experiments, inferences and statements

Experimental and investigative skills

Chemistry is a practical subject and part of the assessment of AS Chemistry will test your practical skills. You will be assessed on three main skills:

- implementing
- analysing evidence and drawing conclusions
- evaluating evidence and procedures.

The skills may be assessed in the context of separate practical exercises, although more than one skill could be assessed in any one exercise. They may also be assessed all together in the context of a single 'whole investigation'. An investigation may be set by your teacher or you may be able to pursue an investigation of your choice. Alternatively, you may take a practical examination.

You will receive guidance about how your practical skills will be assessed from your teacher. This study guide concentrates on preparing you for the written examinations testing the subject content of AS Chemistry.

Different types of questions in AS examinations

In AS Chemistry examinations, different types of question are used to assess your abilities and skills. Unit tests mainly use structured questions requiring both short answers and more extended answers.

Short-answer questions

A short-answer question may test recall or it may test understanding. Short-answer questions normally have space for the answers printed on the question paper.

Here are some examples (the answers are shown in a *handwriting font*):

What is meant by isotopes?

Atoms of the same element with different masses.

Calculate the amount (in mol) of H_2O in 4.5 g of H_2O.

1 mol H_2O has a mass of 18 g. ∴ 4.5 g of H_2O contains 4.5/18 = 0.25 mol H_2O.

Structured questions

Structured questions are in several parts. The parts usually have a common context and they often become progressively more difficult and more demanding as you work your way through the question. A structured question may start with simple recall, then test understanding of a familiar or an unfamiliar situation.

Most of the practice questions in this book are structured questions, as this is the main type of question used in the assessment of AS Chemistry.

When answering structured questions, do not feel that you have to complete one question before starting the next. The further you are into a question, the more difficult the marks are to obtain. If you run out of ideas, go on to the next question. You need to respond to as many parts of questions on an exam paper as possible. You will not score well if you spend so long trying to perfect the first questions that you do not have time to attempt later questions.

Here is an example of a structured question that becomes progressively more demanding.

(a) Write down the atomic structure of the two isotopes of potassium: ^{39}K and ^{41}K.

(i) *^{39}K ...19... protons; ...20... neutrons; ...19... electrons.* ✓

(ii) *^{41}K ...19... protons; ...22... neutrons; ...19... electrons.* ✓ [2]

(b) A sample of potassium has the following percentage composition by mass: ^{39}K: 92%; ^{41}K: 8% Calculate the relative atomic mass of the potassium sample.

92 x 39/100 + 8 x 41/100 = 39.16 ✓ [1]

(c) What is the electronic configuration of a potassium atom?

$1s^2 2s^2 2p^6 3s^2 3p^6 4s^1$ ✓ [1]

(d) The second ionisation energy of potassium is much larger than its first ionisation energy.

(i) Explain what is meant by the *first ionisation energy* of potassium.

The energy required to remove an electron ✓ from each atom in 1 mole ✓ of gaseous atoms ✓

(ii) Why is there a large difference between the values for the first and the second ionisation energies of potassium?

The 2nd electron removed is from a different shell ✓ which is closer to the nucleus and experiences more attraction from the nucleus. ✓ This outermost electron experiences less shielding from the nucleus because there are fewer inner electron shells than for the 1st ionisation energy. ✓ [6]

Extended answers

In AS Chemistry, questions requiring more extended answers may form part of structured questions or may form separate questions. They may appear anywhere on the paper and will typically have between 5 and 10 marks allocated to the answers as well as several lines of answer space. These questions are also often used to assess your abilities to communicate ideas and put together a logical argument.

The correct answers to extended questions are often less well-defined than to those requiring short answers. Examiners may have a list of points for which credit is awarded up to the maximum for the question.

An example of a question requiring an extended answer is shown below.

Magnesium oxide and sulfur trioxide are both solids at 10 °C. Magnesium oxide has a melting point of 2852 °C. Sulfur trioxide has a melting point of 17 °C. Explain, in terms of structure and bonding, why these two compounds have such different melting points. [7]

Points that the examiners might look for include:

Magnesium oxide has a giant lattice structure ✓ containing ionic bonds. ✓ These strong forces need to be broken during melting to allow ions to move free of the rigid lattice. ✓ This requires a large amount of energy supplied by a high temperature. ✓

Sulfur trioxide has a simple molecular structure ✓ held together by van der Waals' forces. ✓ These weak forces need to be broken during melting to allow molecules to move free of the rigid lattice. ✓ This requires a small amount of energy supplied by a low temperature. ✓

8 marking points → [7]

In this type of response, there may be an additional mark for a clear, well-organised answer, using specialist terms. In addition, marks may be allocated for legible text with accurate spelling, punctuation and grammar.

Other types of questions

Free-response and open-ended questions allow you to choose the context and to develop your own ideas. These are little used for assessing AS Chemistry but, if you decide to take the full A Level in Chemistry, you will encounter this type of question during synoptic assessment.

Multiple-choice or objective questions require you select the correct response to the question from a number of given alternatives. These are rarely used for assessing AS Chemistry, although it is possible that some short-answer questions will use a multiple-choice format.

Exam technique

Links from GCSE

Advanced Subsidiary Chemistry builds from grade CC in GCSE Science and GCSE Additional Science (combined), or GCSE Chemistry. This study guide has been written so that you will be able to tackle AS Chemistry from a GCSE Science background.

You should not need to search for important Chemistry from GCSE Science because this has been included where needed in each chapter. If you have not studied Science for some time, you should still be able to learn AS Chemistry using this text alone.

What are examiners looking for?

Examiners use instructions to help you to decide the length and depth of your answer.

If a question does not seem to make sense, you may have misread it – read it again!

State, define or list

This requires a short, concise answer, often recall of material that can be learnt by rote.

Explain, describe or discuss

Some reasoning or some reference to theory is required, depending on the context.

Outline

This implies a short response, almost a list of sentences or bullet points.

Predict or deduce

You are not expected to answer by recall but by making a connection between pieces of information.

Suggest

You are expected to apply your general knowledge to a 'novel' situation, one which you have not directly studied during the AS Chemistry course.

Calculate

This is used when a numerical answer is required. You should always use units in quantities and significant figures should be used with care.

Look to see how many significant figures have been used for quantities and give your answer to this degree of accuracy.

If the question uses three significant figures, then give your answer in three significant figures also.

Some dos and don'ts

Dos

Do answer the question

No credit can be given for good Chemistry that is irrelevant to the question.

Do use the mark allocation to guide how much you write

Two marks are awarded for two valid points – writing more will rarely gain more credit and could mean wasted time or even contradicting earlier valid points.

Do use diagrams, equations and tables in your responses

Even in 'essay-type' questions, these offer an excellent way of communicating chemistry.

Do write legibly

An examiner cannot give marks if the answer cannot be read.

Do write using correct spelling and grammar. Structure longer essays carefully

Marks are now awarded for the quality of your language in exams.

Don'ts

Don't fill up any blank space on a paper

In structured questions, the number of dotted lines should guide the length of your answer.

If you write too much, you waste time and may not finish the exam paper. You also risk contradicting yourself.

Don't write out the question again

This wastes time. The marks are for the answer!

Don't contradict yourself

The examiner cannot be expected to choose which answer is intended. You could lose a hard-earned mark, e.g.

A covalent bond is a shared pair of electrons ✓ bonded by the electrostatic attraction between ions. ✗

Don't spend too much time on a part that you find difficult

You may not have enough time to complete the exam. You can always return to a difficult part if you have time at the end of the exam.

What grade do you want?

Everyone would like to improve their grades but you will only manage this with a lot of hard work and determination. You should have a fair idea of your natural ability and likely grade in Chemistry and the hints below offer advice on improving that grade.

For a Grade A

You will need to be a very good all-rounder.

- You must go into every exam knowing the work extremely well.
- You must be able to apply your knowledge to new, unfamiliar situations.
- You need to have practised many, many exam questions so that you are ready for the type of question that will appear.

The exams test all areas of the syllabus and any weaknesses in your Chemistry will be found out. There must be no holes in your knowledge and understanding. For a Grade A, you must be competent in all areas.

At A level, you also have the opportunity to achieve an A* grade.

For a Grade C

You must have a reasonable grasp of Chemistry but you may have weaknesses in several areas and you may be unsure of some of the reasons for the Chemistry.

- Many Grade C candidates are just as good at answering questions as the Grade A students but holes and weaknesses often show up in just some topics.
- To improve, you will need to master your weaknesses and you must prepare thoroughly for the exam. You must become a better all-rounder.

For a Grade E

You cannot afford to miss the easy marks. Even if you find Chemistry difficult to understand and would be happy with a Grade E, there are plenty of questions in which you can gain marks.

- You must memorise all definitions.
- You must practise exam questions to give yourself confidence that you do know some Chemistry. In exams, answer the parts of questions that you know first. You must not waste time on the difficult parts. You can always go back to these later.
- The areas of Chemistry that you find most difficult are going to be hard to score on in exams. Even in the difficult questions, there are still marks to be gained. Show your working in calculations because credit is given for a sound method. You can always gain some marks if you get part of the way towards the solution.

What marks do you need?

The table below shows how your average mark is transferred into a grade.

average	*80%*	*70%*	*60%*	*50%*	*40%*
grade	*A*	*B*	*C*	*D*	*E*

Introduced for A levels in September 2008, the first A* grades will be awarded in summer 2010.

To achieve an A* grade, you need to achieve a...

- grade A overall (80% or more on uniform mark scale) for the **whole** A level qualification
- grade A* (90% or more on the uniform mark scale) across your A2 units.

A* grades are awarded for the A level qualification only and not for the AS qualification or individual units.

Four steps to successful revision

Step 1: Understand

- Study the topic to be learned slowly. Make sure you understand the logic or important concepts.
- Mark up the text if necessary – underline, highlight and make notes.
- Re-read each paragraph slowly.

GO TO STEP 2

Step 2: Summarise

- Now make your own revision note summary:
  What is the main idea, theme or concept to be learned?
  What are the main points? How does the logic develop?
  Ask questions: Why? How? What next?
- Use bullet points, mind maps, patterned notes.
- Link ideas with mnemonics, mind maps, crazy stories.
- Note the title and date of the revision notes (e.g. Chemistry: Atomic structure, 3rd March).
- Organise your notes carefully and keep them in a file.

This is now in **short-term memory**. You will forget 80% of it if you do not go to Step 3. **GO TO STEP 3**, but first take a 10 minute break.

Step 3: Memorise

- Take 25 minute learning 'bites' with 5 minute breaks.
- After each 5 minute break test yourself:
  Cover the original revision note summary
  Write down the main points
  Speak out loud (record on tape)
  Tell someone else
  Repeat many times.

The material is well on its way to **long-term memory**. You will forget 40% if you do not do step 4. **GO TO STEP 4**

Step 4: Track/Review

- Create a Revision Diary (one A4 page per day).
- Make a revision plan for the topic, e.g. 1 day later, 1 week later, 1 month later.
- Record your revision in your Revision Diary, e.g.
  Chemistry: Atomic Structure, 3rd March 25 minutes
  Chemistry: Atomic Structure, 5th March 15 minutes
  Chemistry: Atomic Structure, 3rd April 15 minutes
  ... and then at monthly intervals.

Chapter 1

Atoms, moles and reactions

The following topics are covered in this chapter:

- *Atoms and isotopes*
- *The electron structure of the atom*
- *Experimental evidence for electron structures*
- *Atomic mass*
- *The mole*
- *Formulae, equations and reacting quantities*
- *Redox reactions*

1.1 Atoms and isotopes

LEARNING SUMMARY

After studying this section you should be able to:

- *recall the relative charge and mass of a proton, neutron and electron*
- *explain the existence of isotopes*
- *use atomic number and mass number to determine atomic structure*

What is an atom?

An atom is the smallest part of an element that can exist on its own. Atoms are so tiny that there are more atoms in a full stop than there are people in the world. It is now possible to see individual atoms by using the most modern and powerful microscopes. However, nobody has yet seen *inside* an atom and we must devise models for the structure of an atom from experimental evidence.

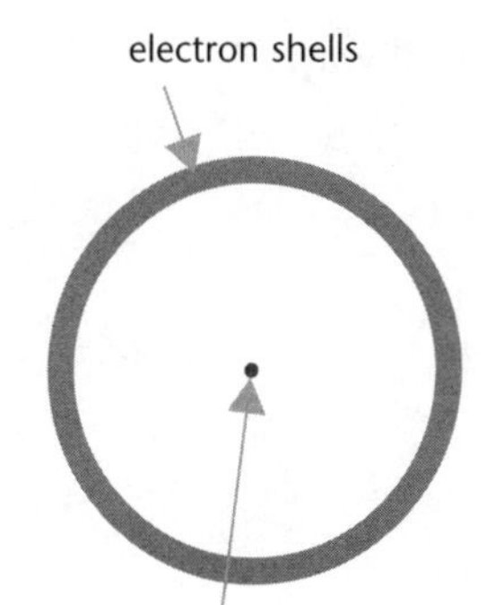

nucleus (protons and neutrons)

Sub-atomic particles

Although there are various models for atomic structure, chemists use a model in which an atom is composed of three **sub-atomic particles**: protons, neutrons and electrons.

In this model, protons and neutrons form the nucleus at the centre of the atom with electrons orbiting in shells. Compared with the total volume of an atom, the nucleus is tiny and extremely dense. Most of an atom is empty, made up of the space between the nucleus and the electron shells.

All matter is made up from the chemical elements and 116 are now known (although others will inevitably be discovered). The number of protons in the nucleus distinguishes the atoms of each element.

An important principle of chemistry is the link between an element and the number of protons in its atoms.

Each atom of an element has the same number of protons.
The number of protons in an atom of an element is called the **atomic number**, Z.

Properties of protons, neutrons and electrons

All atoms of hydrogen contain 1 proton.

All atoms of carbon contain 6 protons.

All atoms of oxygen contain 8 protons.

Some properties of protons, neutrons and electrons are shown in the table below.

particle	*relative mass*	*relative charge*
proton, p	1	1+
neutron, n	1	0
electron, e	1/1840	1–

A proton has virtually the same mass as a neutron and, for most of chemistry, their masses can be assumed to be identical.

All atoms of hydrogen contain:
1 proton and 1 electron.

All atoms of carbon contain:
6 protons and 6 electrons.

All atoms of oxygen contain:
8 protons and 8 electrons.

An electron has negligible mass compared with the mass of a proton or a neutron. In most of chemistry, the mass of an electron can be ignored.

The charge on a proton is opposite to that of an electron but each charge has the same magnitude. An atom is electrically neutral. To balance out the charges, an atom must have the same number of protons as electrons.

> **KEY POINT**
> An atom is electrically neutral.
> An atom contains the same number of protons as electrons.

Isotopes

Without neutrons, the nucleus would just contain positively-charged protons. Like-charges repel, and a nucleus containing just protons would fly apart! Strong nuclear forces act between protons and neutrons and these hold the nucleus together. Neutrons can be thought of as 'nuclear glue'.

In each atom of an element, the number of protons in the nucleus is fixed. However, most elements contain atoms with different numbers of neutrons. These atoms are called **isotopes** and, because they have different numbers of neutrons, they also have different masses.

> **KEY POINT**
> Isotopes are atoms of the **same** element with **different** masses.
> Isotopes of an element have the **same** number of **protons and electrons**.
> Each isotope has a **different** number of **neutrons** in the nucleus.
> The combined number of protons and neutrons in an isotope of an element is called the **mass number**, *A*.

The isotopes of an element react in the same way. This is because chemical reactions involve electrons – neutrons make no difference.

Chemists represent the nucleus of an isotope in a special way.

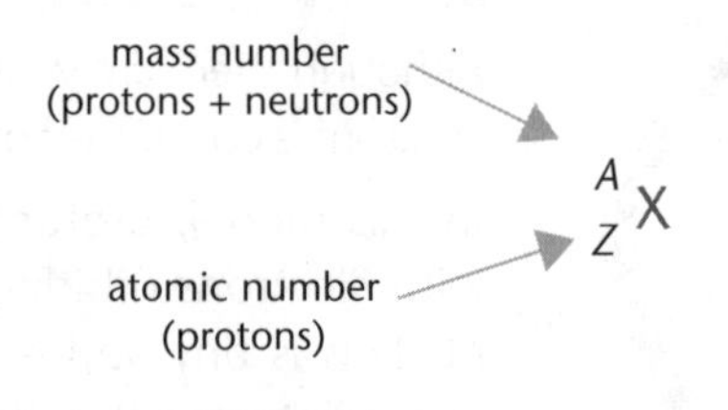

The isotopes of carbon

Carbon exists as three isotopes, $^{12}_{6}C$, $^{13}_{6}C$, $^{14}_{6}C$. In all but the most accurate work, it is reasonable to assume that the relative mass of an isotope is equal to its mass number. It is easy to work out the atomic structure of an isotope from the representation above.

isotope	*atomic number*	*mass number*	*protons*	*neutrons*	*electrons*
$^{12}_{6}C$	6	12	6	6	6
$^{13}_{6}C$	6	13	6	7	6
$^{14}_{6}C$	6	14	6	8	6

Because all carbon isotopes have 6 protons, it is common practice to omit the atomic number, and $^{12}_{6}C$ is often shown as ^{12}C, or even as carbon-12.

Progress check

How many protons, neutrons and electrons are in the following isotopes?

1 $^{7}_{3}Li$; **2** $^{23}_{11}Na$; **3** $^{19}_{9}F$; **4** $^{27}_{13}Al$; **5** $^{55}_{26}Fe$.

1 $^{7}_{3}Li$: 3p, 4n, 3e
2 $^{23}_{11}Na$: 11p, 12n, 11e
3 $^{19}_{9}F$: 9p, 10n, 9e;
4 $^{27}_{13}Al$: 13p, 14n, 13e;
5 $^{55}_{26}Fe$: 26p, 29n, 26e.

1.2 The electron structure of the atom

After studying this section you should be able to:

- *understand the relationship between shells, sub-shells and orbitals*
- *understand how atomic orbitals are filled*
- *describe electron structure in terms of shells and sub-shells*

Energy levels or 'shells'

AQA M1

You can imagine a model of an atom with electrons orbiting in shells around the nucleus. The electrons in each successive shell have an orbit further away from the nucleus. The further a shell is from the nucleus, the greater the shell's energy level. Each energy level is given a number called the principal quantum number, *n*. The shell closest to the nucleus has an energy level with $n = 1$, then $n = 2$ and so on.

energy

$n = 4$ — $32e^-$

$n = 3$ — $18e^-$

$n = 2$ — $8e^-$

$n = 1$ — $2e^-$

Electron energy levels are like a ladder and are filled from the bottom up.

Notice that the gap between successive energy levels becomes less with increasing energy.

The number of electrons that can occupy the first four energy levels is shown below:

n	*shell*	*electrons*
1	1st shell	2
2	2nd shell	8
3	3rd shell	18
4	4th shell	32

The principal quantum shell, *n*, is the shell number.

If you look closely, you can see a pattern. The number of electrons that can occupy a shell is $2n^2$.

Using this model, electrons can be placed into available shells, starting with the lowest energy level. Each shell must be full before the next starts to fill. The table below shows how the shells are filled for the first 11 elements in the Periodic Table.

element	*atomic number*	*electrons* $n = 1$	$n = 2$	$n = 3$
H	1	1		
He	2	2		
Li	3	2	1	
Be	4	2	2	
B	5	2	3	
C	6	2	4	
N	7	2	5	
O	8	2	6	
F	9	2	7	
Ne	10	2	8	
Na	11	2	8	1

This model breaks down as the $n = 3$ energy level is filled because each shell consists of sub-shells.

Sub-shells and orbitals

AQA M1

A more advanced model of electron structure is used in which each shell is made up of **sub-shells**.

Sub-shells

There are different types of sub-shell: s, p, d and f. Each type of sub-shell can hold a different number of electrons:

sub-shell	*electrons*
s	2
p	6
d	10
f	14

The table below shows the shells and sub-shells for the first four principal quantum numbers.

Notice the labelling. The s sub-shell in the 2nd shell is labelled 2s.

n	*shell*	*sub-shell*				*total number of electrons*	
1	1st shell	1s				2	= 2
2	2nd shell	2s	2p			2 + 6	= 8
3	3rd shell	3s	3p	3d		2 + 6 + 10	= 18
4	4th shell	4s	4p	4d	4f	2 + 6 + 10 + 14	= 32

- Each successive shell contains a new type of sub-shell.
- The 1st shell contains 1 sub-shell, the second shell contains 2 sub-shells, and so on.

Orbitals

How do the electrons fit into the sub-shells? Mathematicians have worked out that electrons occupy negative charge clouds called **orbitals** and these make up each sub-shell.

KEY POINT

- An orbital can hold up to two electrons.
- Each type of sub-shell has different orbitals: s, p, d and f.

The table below shows how electrons fill the orbitals in each sub-shell.

sub-shell	*orbitals*	*electrons*
s	1	1 x 2 = 2
p	3	3 x 2 = 6
d	5	5 x 2 = 10
f	7	7 x 2 = 14

s-orbitals

One s-orbital

- An s-orbital has a spherical shape.

p-orbitals

Three p-orbitals

- A p-orbital has a 3-dimensional dumb-bell shape.
- There are three p-orbitals, p_x, p_y and p_z, at right angles to one another.

Three p-orbitals

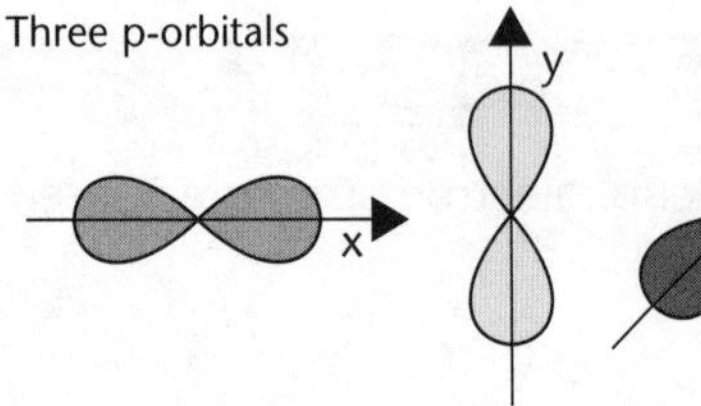

giving

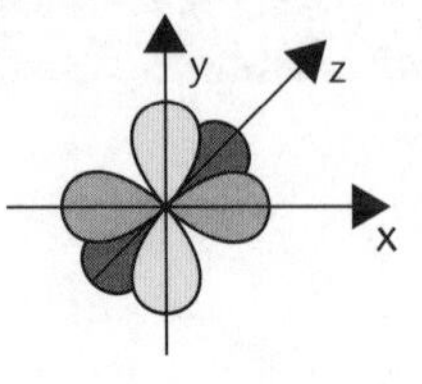

Orbitals are regions around a nucleus that have electron density.

Five d-orbitals
Seven f-orbitals

d-orbitals and f-orbitals

The structures of d and f-orbitals are more complex.

- There are five d-orbitals.
- There are seven f-orbitals.

How do two electrons fit into an orbital?

'Electrons in a box'
allowed

not allowed

Electrons are negatively charged and so they repel one another. An electron also has a property called **spin**. The two electrons in an orbital have opposite spins, helping to counteract the natural repulsion between their negative charges.

> **KEY POINT**
> - Chemists often represent an orbital as a box that can hold up to 2 electrons.
> - Each electron is shown as an arrow, indicating its spin: either ↑ or ↓.
> - Within an orbital, the electrons must have **opposite spins**.

Filling the sub-shells

AQA M1

Sub-shells have different energy levels. The diagram below shows the relative energies for the sub-shells in the first four shells.

Within a shell, the sub-shell energies are in the order: s, p, d and f.

Note that the 4s sub-shell is at a lower energy than the 3d sub-shell.

The 4s sub-shell fills before the 3d sub-shell.

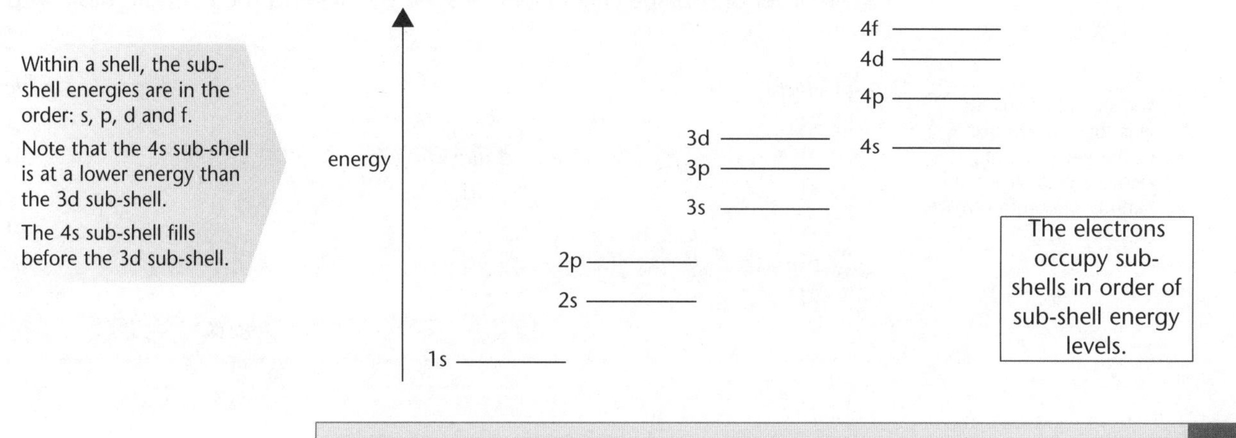

> **KEY POINT**
> - Shells and sub-shells are occupied in energy-level order.
> - Electrons occupy orbitals singly before pairing begins to prevent any repulsion caused by pairing.

The 2s orbital is occupied before the 2p orbitals because it is at a lower energy.

Note that the 2p orbitals are occupied singly before pairing begins.

The diagram below shows how electrons occupy orbitals from boron to oxygen.

Use these 4 examples to see how electron configuration is written.

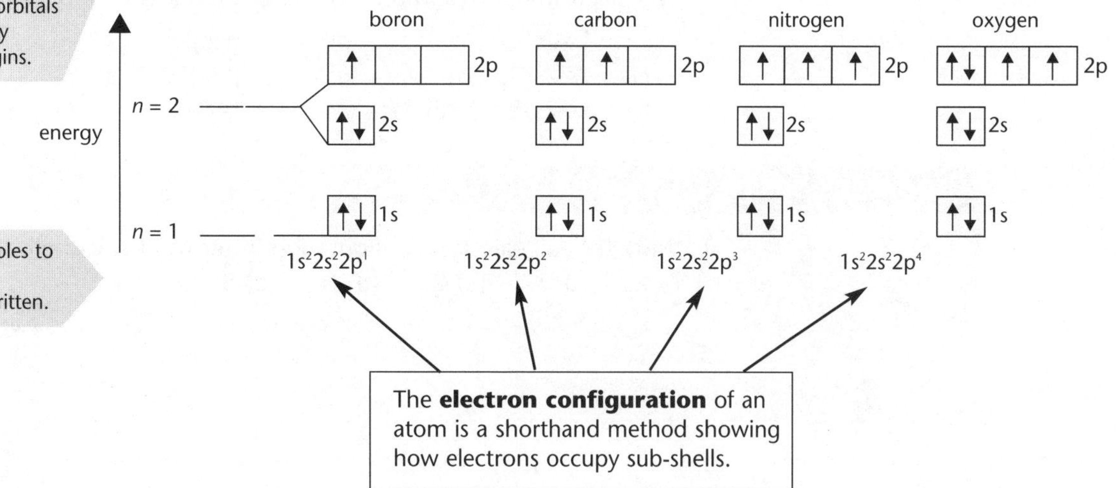

The diagram below shows how the sub-shells are filled in an atom of potassium. Notice why the 4s sub-shell starts to fill before the 3d sub-shell.

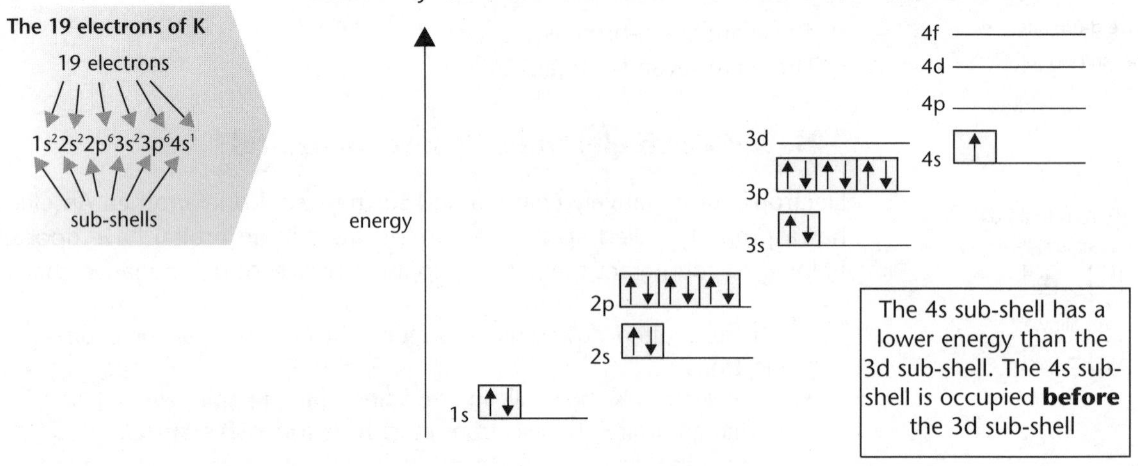

Sub-shells and the Periodic Table

AQA M1

The Periodic Table is structured in blocks of 2, 6, 10 and 14, linked to sub-shells. The order of sub-shell filling can be seen by dividing the Periodic Table into blocks.

You should be able to work out the electron configuration for any element in the first four periods, i.e. up to krypton ($Z = 36$).

s-block	*d-block*	*p-block*
1s		1s
2s		2p
3s		3p
4s	3d	4p
5s	4d	5p
6s	5d	6p
7s		

f-block
4f
5f

Simplifying electron configurations

The similar electron configurations within a group of the Periodic Table can be emphasised with a simpler representation in terms of the previous noble gas.

The electron configurations for the elements in Group 1 are shown below:

Li: $1s^2 2s^1$ *or* $[He]2s^1$
Na: $1s^2 2s^2 2p^6 3s^1$ *or* $[Ne]3s^1$
K: $1s^2 2s^2 2p^6 3s^2 3p^6 4s^1$ *or* $[Ar]4s^1$

Progress check

1 Write the full electron configuration in terms of sub-shells for:
(a) C; (b) Al; (c) Ca; (d) Fe; (e) Br.

1 (a) C: $1s^2 2s^2 2p^2$
(b) Al: $1s^2 2s^2 2p^6 3s^2 3p^1$
(c) Ca: $1s^2 2s^2 2p^6 3s^2 3p^6 4s^2$
(d) Fe: $1s^2 2s^2 2p^6 3s^2 3p^6 3d^6 4s^2$
(e) Br: $1s^2 2s^2 2p^6 3s^2 3p^6 3d^{10} 4s^2 4p^5$

1.3 Experimental evidence for electron structures

After studying this section you should be able to:

LEARNING SUMMARY

- *recall the definitions for first and successive ionisation energies*
- *understand the factors affecting the sizes of ionisation energies*
- *show how ionisation energies provide evidence for sub-shells*
- *understand evidence for electron structure from electron emission and absorption spectra*

Ions

AQA M1

An atom can either lose or gain electrons to form an **ion**: a charged atom.

A positive ion is formed when an atom loses electrons. For example, a lithium atom forms a **positive ion** with a 1+ charge by **losing** an electron:

$$\underset{(3p^+,\ 4n,\ \mathbf{3e^-})}{Li} \longrightarrow \underset{(3p^+,\ 4n,\ \mathbf{2e^-})}{Li^+} + e^-$$

Positive ions form when electrons are lost.

Negative ions form when electrons are gained.

A negative ion is formed when an atom gains electrons. For example, an oxygen atom forms a **negative ion** with a 2– charge by **gaining** two electrons:

$$\underset{(6p^+,\ 6n,\ \mathbf{6e^-})}{O} + 2e^- \longrightarrow \underset{(6p^+,\ 6n,\ \mathbf{8e^-})}{O^{2-}}$$

Ionisation energy

AQA M1

Ionisation energy measures the ease with which electrons are lost in the formation of positive ions. An element has as many ionisation energies as there are electrons.

KEY POINT

The **first** *ionisation energy* of an element is the energy required to remove **1 electron** from each atom in **1 mole** of **gaseous atoms** to form 1 mole of gaseous 1+ ions.

The equation representing the first ionisation energy of sodium is shown below.

$Na(g) \longrightarrow Na^+(g) + e^-$ 1st ionisation energy = +496 kJ mol^{-1}

Factors affecting ionisation energy

Electrons are held in their shells by attraction from the nucleus. The first electron lost will be from the highest occupied energy level. This electron experiences least attraction from the nucleus.

Three factors affecting the size of this attraction are shown below.

These factors are very important and help to explain many chemical ideas throughout the course.

Learn them!

KEY POINT

Atomic radius

The greater the distance between the nucleus and the outer electrons, the less the attractive force. Attraction falls rapidly with increasing distance and so this factor is very important and has a big effect.

Nuclear charge

The greater the number of protons in the nucleus, the greater the attractive force.

Electron shielding or 'screening'

The outer shell electrons are repelled by any inner shells between the electrons and the nucleus.

This repelling effect, called electron shielding or screening, reduces the overall attractive force experienced by the outer electrons.

Trends in first ionisation energies

AQA M1

The graph below shows the variation of first ionisation energy with increasing atomic number from hydrogen to calcium:

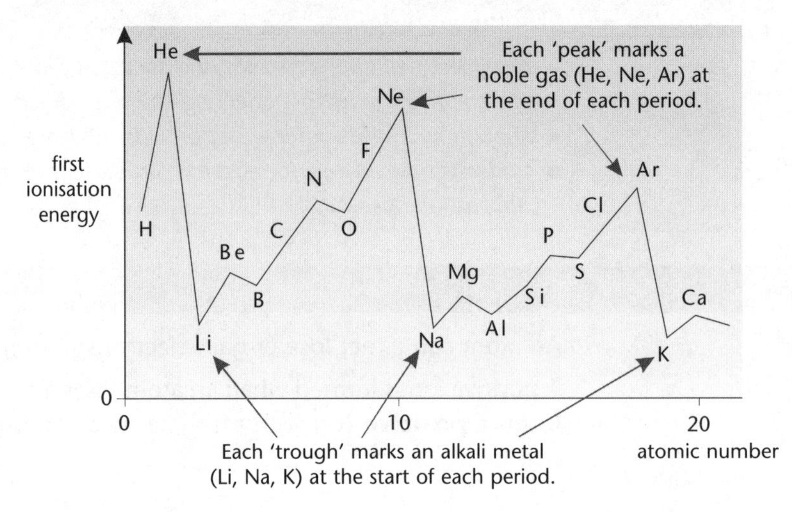

The graph shows:

Across a period, increased nuclear charge is most important.

- a general **increase** in first ionisation energy across a period (see H→He; Li→Ne; Na→Ar).
  This results from the **increase** in nuclear charge as electrons are added to the **same shell** across each period.
- a sharp **decrease** in first ionisation energy between the end of one period and the start of the next period (see He→Li; Ne→Na; Ar→K).
  This reflects the addition of a new outer shell with the resulting increase in **distance** and **shielding**.

Down a group, increased distance and shielding are most important.

- a **decrease** in first ionisation energy down a group (see He→Ne→Ar and other groups).
  This reflects the presence of **extra shells** down the group.

The reasons for the general trend in ionisation energies are similar to those explaining the trend in atomic radii (page 57).

Evidence for sub-shells

AQA M1

The variation in first ionisation energies across a period in the Periodic Table provides evidence for the existence of sub-shells. The graph below shows this variation across Period 2 (Li → Ne)

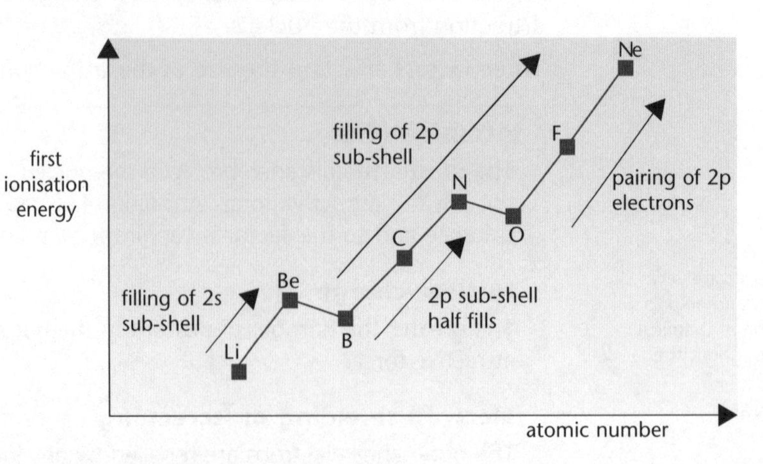

Across the period, the graph shows:

- a rise from lithium to beryllium,
- a fall to boron followed by a rise to nitrogen,
- a fall to oxygen followed by a rise to neon.

Comparing beryllium and boron

The fall in 1st ionisation energy from beryllium to boron marks the start of filling the 2p sub-shell which is at higher energy than the 2s sub-shell.

Beryllium versus boron: higher energy level.

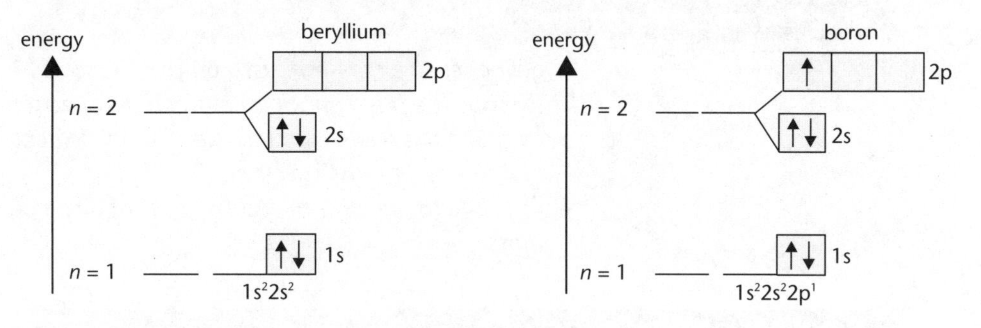

- Beryllium's outermost electron is in the 2s sub-shell.
- Boron's outermost electron is in the 2p sub-shell.
- Boron's outermost electron is easier to remove because the 2p sub-shell has a **higher energy level** than the 2s sub-shell.

Comparing nitrogen and oxygen

The fall in 1st ionisation energy from nitrogen to oxygen marks the start of electron pairing in the p-orbitals of the 2p sub-shell.

Nitrogen versus oxygen: electron pairing.

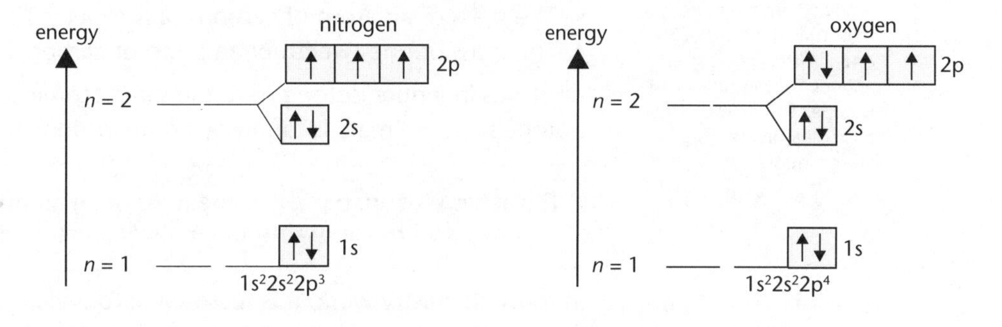

- Both nitrogen and oxygen have their outer electrons in the same 2p sub-shell with the same energy level.
- Oxygen has one 2p orbital with an electron pair.
- Nitrogen, with one electron in each 2p orbital, has unpaired 2p electrons only.
- Because of **electron repulsion**, it is easier to remove one of oxygen's paired 2p electrons.

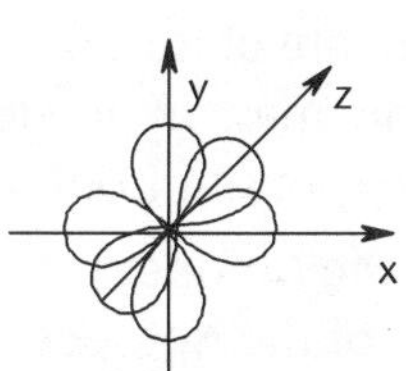

Nitrogen
2p sub-shell half-full
1 electron in each
2p orbital parallel
spins at right angles

y
z
x

Oxygen
2p electrons
start to pair
Paired
electrons repel

Progress check

1 State and explain the general trend in 1st ionisation energy across a period.

2 Explain why the 1st ionisation energy falls slightly:
(a) between beryllium and boron
(b) between nitrogen and oxygen.

1 The 1st ionisation energy decreases across a period. Across a period the number of protons increases and electrons are being added to the same shell. Therefore attraction from the nucleus increases.

2 (a) In B, the highest energy electron is in the 2p sub-shell, higher in energy than the 2s sub-shell in Be.
(b) In O, 2 electrons pair up in a 2p sub-shell. In N, the 2p sub-shell is half filled with a single electron in each orbital. The paired electrons in O repel one another slightly making it easier to remove an electron.

1.4 Atomic mass

After studying this section you should be able to:

- *define relative masses, based on the ^{12}C scale*
- *describe the basic principles of the mass spectrometer*
- *calculate the relative atomic mass of an element from its isotopic abundance or mass spectrum*
- *calculate the relative molecular mass of a compound from relative atomic masses*

LEARNING SUMMARY

Relative atomic mass

AQA M1

Carbon-12: an international standard

The mass of an atom is too small to be measured on even the most sensitive balance. Chemists use **relative** masses to compare the atomic masses of different elements. First, we must have an atom with which to compare other atoms and an atom of the **carbon-12 isotope** is chosen as the international standard for the measurement of atomic mass.

Unified atomic mass unit (u)

1 unified atomic mass unit (u) is the mass of one-twelfth the mass of an atom of the carbon-12 isotope. This provides the base measurement for atomic masses:

1 u = 1.661×10^{-23} g

- The mass of an atom of carbon-12 is exactly 12 unified atomic mass units (u).
- The mass of one-twelfth of an atom of carbon-12 is exactly 1 u.

All atoms in a pure isotope have the same atomic structure and the same mass. An isotope's *relative* mass is found by comparison with carbon-12.

KEY POINT

Relative isotopic mass is the mass of an atom of an isotope compared with one-twelfth the mass of an atom of carbon-12.

In most chemistry work, it is reasonable to:

- neglect the tiny contribution to atomic mass from electrons
- take both the mass of a proton and a neutron as 1 u.

Relative isotopic mass is then simply the mass number of the isotope, e.g. the relative isotopic mass of ^{16}O is 16.

Relative atomic mass

Note that a relative mass has no units. It is simply a ratio of masses and any mass units will cancel.

Most elements consist of a mixture of isotopes, each with a different mass number. To work out the relative atomic mass of an element we must find the **weighted average mass** of the isotopes present from:

- the natural abundances of the isotopes
- the relative isotopic masses of the isotopes.

Examples of relative atomic masses:

H: 1.008

Cl: 35.45

Pb: 207.19

KEY POINT

Relative atomic mass, A_r, is the weighted average mass of an atom of an element compared with one-twelfth of the mass of an atom of carbon-12.

Calculating a relative atomic mass

In most chemistry work, the relative isotopic mass can be taken as a whole number.

Naturally occurring chlorine consists of 75% ^{35}Cl and 25% ^{37}Cl.

$\frac{75}{100}$ of chlorine is the ^{35}Cl isotope;

$\frac{25}{100}$ of chlorine is the ^{37}Cl isotope.

The relative atomic mass of chlorine = $\frac{75}{100} \times 35 + \frac{25}{100} \times 37 = 35.5$

Measuring relative atomic masses

A relative atomic mass can be determined using a **mass spectrometer**.

Remember VIADD:
Vaporisation
Ionisation
Acceleration
Deflection
Detection

- A sample of the element is placed into the mass spectrometer and vaporised.
- The sample is bombarded with electrons forming positive ions.
- The positive ions are accelerated using an electric field.
- The positive ions are deflected using a magnetic field.
- Ions of lighter isotopes are deflected more than ions of heavier isotopes. This separates different isotopes.
- The ions are detected to produce a **mass spectrum**.

Notice that ions are detected in the mass spectrometer. The ions are formed following electron bombardment.

You can follow the stages in the diagram below:

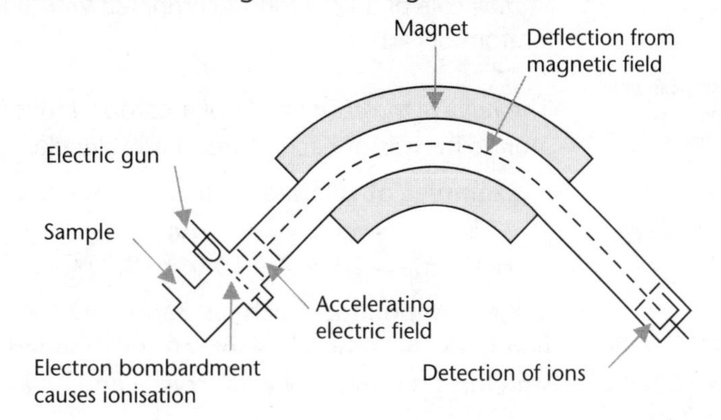

Relative atomic mass from a mass spectrum

A mass spectrum provides:

- the relative isotopic masses of the isotopes in an element
- isotopic abundances.

The diagram below shows how a relative atomic mass can be calculated from a mass spectrum.

Mass spectrum of a copper sample

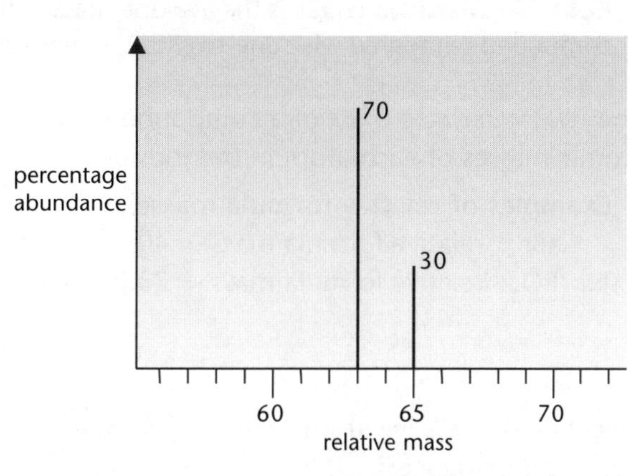

ion	*relative mass*	*percentage abundance*
$^{63}Cu^{+}$	63	70%
$^{65}Cu^{+}$	65	30%

Mass spectrometers have been sent into space to identify elements, e.g. Mars space probe. On Earth, mass spectrometers are used in environmental monitoring and for forensic analysis.

The relative atomic mass of copper

$$= \frac{70}{100} \times 63 + \frac{30}{100} \times 65$$

$$= 63.6$$

More relative masses

AQA M1

Relative masses can be used to compare the masses of any chemical species. Much of matter is made up of compounds.

Relative molecular mass

You will find out more about chemical bonding and structure in Chapter 2, p.39.

A simple molecule of Cl_2, H_2O or CO_2 comprises small groups of atoms, held together by chemical bonds.

> **Relative molecular mass**, M_r, is the weighted average mass of a molecule of a compound compared with one-twelfth the mass of an atom of carbon-12.
>
> KEY POINT

Mass spectrometry can be used with compounds to determine the relative molecular mass.

The *relative molecular mass* of a compound is found by adding together the relative atomic masses of each atom in a molecule.

Examples of relative molecular masses

Cl_2: $M_r = 35.5 \times 2 = 71.0$;

H_2O: $M_r = 1.0 \times 2 + 16.0 = 18.0$

For giant structures, the **formula unit** of the compound is the simplest ratio of atoms or ions in the structure.

Some compounds, such as sand SiO_2, exist as 'giant' molecules comprising hundreds of thousands of atoms bonded together. Each molecule in these structures is the size of each crystal and, in a sense, the whole crystal is a molecule.

- Each 'molecule' of SiO_2 has twice as many oxygen atoms as silicon atoms.

Many compounds, such as common salt NaCl, are made up of ions and not molecules. These ionic compounds do not have single molecules and form giant structures of oppositely charged ions.

- NaCl has the same number of sodium ions as chloride ions.

Relative formula mass can be used to describe the relative mass of any compound.

Although relative **molecular** mass is often used with these compounds, a better term to use is relative **formula** mass.

> **Relative formula mass** is the average mass of the formula unit of a compound compared with one-twelfth the mass of an atom of carbon-12.
>
> KEY POINT

The *relative formula mass* of a compound is found by adding together the relative atomic masses of each atom in the formula.

Examples of relative formula masses

$CaBr_2$: relative formula mass $= 40.1 + 79.9 \times 2 = 199.9$;

Na_3PO_4: relative formula mass $= 23.0 \times 3 + 31.0 + 16.0 \times 4 = 164.0$

Progress check

1 Calculate the relative atomic mass, A_r, of the following elements from their isotopic abundances:

(a) Gallium containing 60% ^{69}Ga and 40% ^{71}Ga.

(b) Silicon containing 92.18% ^{28}Si, 4.71% ^{29}Si and 3.11% ^{30}Si.

2 Use A_r values from the Periodic Table to calculate the relative formula mass of:

(a) CO_2

(b) NH_3

(c) Fe_2O_3

(d) $C_6H_{12}O_6$

(e) $Pb(NO_3)_2$.

1 (a) Gallium: 69.8
(b) Silicon: 28.11.

2 (a) 44.0
(b) 17.0
(c) 159.6
(d) 180.0
(e) 331.2.

1.5 The mole

After studying this section you should be able to:

LEARNING SUMMARY

- *understand the concept of the mole as the amount of substance*
- *calculate molar quantities using masses, solutions and gas volumes*

The mole

AQA M1

Counting and weighing atoms

To make a compound such as H_2O, it would be useful to be able to count out atoms. For H_2O, we would need twice as many hydrogen atoms as oxygen atoms. Atoms are far too small to be counted individually but they can be counted using the idea of relative mass:

For 1 atom of H, C and O, the relative masses are in the ratio 1 : 12 : 16.
For 10 atoms of each, the ratio is 10 : 120 : 160 which is still 1 : 12 : 16.

Provided we have the same number of atoms of each, the masses are in the ratio of the relative masses.

By measuring masses, chemists are able to actually count atoms.
For 1 g of H, 12 g of C and 16 g of O, the masses are still in the ratio 1 : 12 : 16.
1 g of H contains the same number of atoms as 12 g of C and 16 g of O.

Amount of substance

Chemists use a quantity called *amount of substance* for counting atoms.
Amount of substance is:

- given the symbol *n*
- measured using a unit called the **mole** (abbreviated as **mol**).

Note that carbon-12 is the **substance** here. So **amount of** ***substance*** translates to **amount of carbon-12.**

As with relative masses, amount of substance uses the carbon-12 isotope as the standard:

The amount *n* of carbon-12 atoms in 12 g of ^{12}C is 1 *mol*.

The actual **number** of atoms in 1 mol of carbon-12 is 6.02×10^{23}.

This can be expressed by saying:

'There are 6.02×10^{23} atoms per mole of carbon-12 atoms (or 6.02×10^{23} mol^{-1})'.

6.02×10^{23} mol^{-1} is a constant known as the **Avogadro constant**, *L*, or N_A.

KEY POINT

This is a very powerful and useful idea. To make a water molecule from hydrogen and oxygen, we cannot **count** out the atoms but we can **weigh** them out in the correct ratio. To make H_2O, we need 2 atoms of hydrogen for every 1 atom of oxygen, which we can get by weighing out 2 g of hydrogen and 16 g of oxygen.

2 g of H contains twice as many atoms as 16 g of O.
2 g of H atoms contains 2 mol of H atoms.
16 g of O atoms contains 1 mol of O atoms.

Particle

A single particle of a substance may refer to an atom, a molecule, an ion, an electron or to any identifiable particle. Chemists refer to a collection of particles as a **chemical species.**

Amount of substance is not restricted just to atoms. We can have an amount of any chemical species: 'amount of atoms', 'amount of molecules', 'amount of ions', etc.

> A **mole** is the amount of substance that contains as many single particles as there are atoms in exactly 12 g of the carbon-12 isotope.
>
> KEY POINT

Using the definition of the mole above, this means that:

- 1 mole of hydrogen atoms, H, contains 6.02×10^{23} hydrogen atoms, H.
- 1 mole of oxygen molecules, O_2, contains 6.02×10^{23} oxygen molecules, O_2.
- 1 mole of electrons, e^-, contains 6.02×10^{23} electrons, e^-.

It is important to always refer to what the particle is.

For example:

'1 mole of oxygen' could refer to oxygen atoms, O, or to oxygen molecules, O_2.

It is always safest to quote the formula to which the amount of substance refers.

Moles from masses

AQA M1

Molar mass, *M*

> **Molar mass, *M***, is the mass per mole of a substance.
> The units of molar mass are g mol^{-1}.
>
> KEY POINT

For the atoms of an element, the molar mass is simply the relative atomic mass in g mol^{-1}.

- Molar mass of Mg = 24.3 g mol^{-1}
- This simply means that each mole of Mg atoms has a mass of 24.3 g.

Molar mass is an extremely useful term that can be applied to any chemical species.

For a compound, the molar mass is found by adding together the relative atomic masses of each atom in the formula:

- Molar mass of CO_2 = 12 + (16 x 2) = 44 g mol^{-1}
- 1 mole of CO_2 has a mass of 44 g.

Examples of molar masses

H	1 g mol^{-1}
H_2O	18 g mol^{-1}
CO_2	44 g mol^{-1}
HNO_3	63 g mol^{-1}

> Amount of substance *n*, mass and molar mass are linked by the expression below:
>
> $$n = \frac{\text{mass (in g)}}{\text{molar mass}}$$
>
> KEY POINT

In 24 g of C, *amount of C* $= \frac{\text{mass (in g)}}{\text{molar mass}} = \frac{24}{12} = 2$ mol

In 11 g of CO_2, *amount of* $CO_2 = \frac{\text{mass (in g)}}{\text{molar mass}} = \frac{11}{44} = 0.25$ mol

Progress check

1 (a) What is the amount of each substance (in mol) in the following:
(i) 124 g P; (ii) 64 g O; (iii) 64 g O_2; (iv) 10.01 g $CaCO_3$; (v) 3.04 g Cr_2O_3?

(b) What is the mass of each substance (in g) in the following:
(i) 0.1 mol Na; (ii) 2.5 mol NO_2; (iii) 0.3 mol Na_2SO_4; (iv) 0.05 mol H_2SO_4; (v) 0.025 mol Ag_2CO_3?

1 (a) (i) 4 mol; (ii) 4 mol; (iii) 2 mol; (iv) 0.1 mol; (v) 0.02 mol.
(b) (i) 2.3 g; (ii) 115 g; (iii) 42.63 g; (iv) 4.905 g; (v) 6.895 g.

Moles from solutions

AQA M1

> **KEY POINT**
> The concentration of a solution is the amount of solute, in mol, dissolved in each dm^3 (1000 cm^3) of solution.

'Dilute' solutions contain few moles of solute in a volume.

If the volume of the solution is in dm^3:

$$n = c \times V \textbf{ (in dm}^3\textbf{)}$$

$n =$ amount of substance, in mol.

$c =$ concentration of solution, in mol dm^{-3}.

$V =$ volume of solution, in dm^3.

Dividing a volume measured in cm^3 by 1000 converts the volume automatically to dm^3.

It is more convenient to measure smaller volumes of solutions in cm^3 and the expression becomes:

$$n = c \times \frac{V \text{ (in cm}^3\text{)}}{1000}$$

Molarity

Molarity refers to the concentration in mol dm^{-3}.

Thus 2 mol dm^3 and 2 Molar (or 2M) mean the same: 2 moles of solute in 1 dm^3 of solution.

Example

What is the amount of NaCl (in mol) in 25.0 cm^3 of an aqueous solution of concentration 2.00 mol dm^{-3}?

$$n(\text{NaCl}) = c \times \frac{V}{1000} = 2.00 \times \frac{25.0}{1000} = 0.0500 \text{ mol}$$

Standard solutions

Concentrations are sometimes referred to in g dm^{-3}.

The Na_2CO_3 here has a **molar** concentration of 0.100 mol dm^{-3} and a **mass** concentration of 10.6 g dm^{-3} of Na_2CO_3.

Chemists often need to prepare **standard solutions** with an exact concentration. Using an understanding of the mole, the mass required to prepare such a solution can easily be worked out.

Example

Find the mass of sodium carbonate required to prepare 250 cm^3 of a 0.100 mol dm^{-3} solution.

Find the amount of Na_2CO_3 (in mol) required in solution:

Very dilute solutions are measured in parts per million (ppm), e.g. ppm is 1 milligram (0.001 g) in 1 dm^3 of solution.

$$\text{amount, } n \text{, of } Na_2CO_3 = c \times \frac{V}{1000} = 0.100 \times \frac{250}{1000} = 0.0250 \text{ mol}$$

Convert moles to grammes

molar mass of Na_2CO_3 = 23.0 × 2 + 12.0 + 16.0 × 3 = 106.0 g mol^{-1}

$$n = \frac{\text{mass}}{\text{molar mass}} \quad \therefore \text{ mass} = n \times \text{molar mass}$$

mass of Na_2CO_3 required = 0.0250 × 106.0 = 2.65 g

Progress check

1 (a) What is the amount of each substance, in mol, in:

(i) 250 cm^3 of a 1.00 mol dm^{-3} solution;

(ii) 10 cm^3 of a 2.0 mol dm^{-3} solution?

(b) Find the concentration, in mol dm^{-3}, for:

(i) 4 mol in 2 dm^3 of solution;

(ii) 0.0100 mol in 100 cm^3 of solution.

(c) Find the concentration, in g dm^{-3}, for:

(i) 2 mol of NaOH in 4 dm^3 of solution;

(ii) 0.500 mol of HNO_3 in 200 cm^3 of solution.

1 (a) (i) 0.25 mol; (ii) 0.02 mol.
(b) (i) 2 mol dm^{-3}; (ii) 0.1 mol dm^{-3}.
(c) (i) 20 g dm^{-3}; (ii) 157.5 g dm^{-3}.

Moles from gas volumes

AQA M1

For a gas, the amount of gas molecules (in mol) is most conveniently obtained by measuring the gas volume.

Provided that the pressure and temperature are the same, equal volumes of gases contain the same number of molecules.

This means that it does not matter which gas is being measured. By measuring the volume, we are indirectly also counting the number of molecules.

This is sometimes summarised by Avogadro's hypothesis:

'Equal volumes of gases contain the same number of molecules under the same conditions of temperature and pressure.'

> **KEY POINT**
> At room temperature and pressure (r.t.p), 298 K (25°C) and 100 kPa, 1 mol of a gas occupies approximately 24 dm^3 = 24 000 cm^3.

At r.t.p., if the volume of the gas is in **dm^3**: $n = \frac{V\text{ (in dm}^3)}{24}$

if the volume of the gas is in **cm^3**: $n = \frac{V\text{ (in cm}^3)}{24000}$

Example

How many moles of gas molecules are in 480 cm^3 of a gas at r.t.p.?

$$n = \frac{V\text{ (in cm}^3)}{24000} = \frac{480}{24000} = 0.0200 \text{ mol of gas molecules.}$$

The Ideal Gas Equation

The conditions will not always be room temperature and pressure. A gas volume depends on temperature and pressure.

The Ideal Gas Equation can be used to find the amount of gas molecules n (in mol) in a volume V (in m^3) at any temperature T (in K) and pressure p (in Pa).

The Ideal Gas Equation: $pV=nRT$ (R = 8.31 J K^{-1} mol^{-1})

Before using $pV = nRT$, you must remember to convert any values to Pa, K and m^3.

Conversion rules

cm^3 to m^3	$\times 10^{-6}$
dm^3 to m^3	$\times 10^{-3}$
°C to K	+ 273
kPa to Pa	$\times 10^3$

Example

A 0.215 g sample of a volatile liquid, **X**, produces 77.5 cm^3 of gas at 100°C and 100 kPa. Calculate the relative molecular mass of **X**.

Use The Ideal Gas Equation:

$$pV = nRT \qquad \therefore n = \frac{pV}{RT}$$

$$\therefore n = \frac{1.00 \times 10^5 \times 7.75 \times 10^{-5}}{8.31 \times 373} = 0.00250 \text{ mol}$$

Find the molar mass:

$$n = \frac{\text{mass}}{\text{molar mass}}$$

$$\therefore \text{molar mass} = \frac{\text{mass}}{n} = \frac{0.215}{0.00250} = 86.0 \text{ g mol}^{-1}$$

$\therefore$ Relative molecular mass M_r = 86.0

Note that M (molar mass) has units: g mol^{-1}

M_r (relative molecular mass) has no units.

Progress check

1 (a) What is the volume, at 25°C and 100 kPa of
(i) 44 g $CO_2(g)$; (ii) 7 g $N_2(g)$; (iii) 5.1 g $NH_3(g)$?

(b) What is the mass, at 25°C and 100 kPa of
(i) 1.2 dm^3 O_2; (ii) 720 cm^3 $CO_2(g)$; (iii) 48 cm^3 $CH_4(g)$?

1 (a) (i) 25 dm^3; (ii) 6.2 dm^3; (iii) 7.4 dm^3.
(b) (i) 1.55 g; (ii) 1.28 g; (iii) 0.031 g.

1.6 Formulae, equations and reacting quantities

After studying this section you should be able to:

- *understand what is meant by empirical and molecular formula*
- *calculate the empirical and molecular formula of a substance*
- *write balanced equations*
- *use chemical equations to calculate reacting masses and reacting volumes (and vice versa)*
- *perform calculations involving volumes and concentrations of solutions in simple acid-base titrations*

LEARNING SUMMARY

Types of chemical formula

AQA M1

Empirical and molecular formula

The **empirical formula** of ethane is CH_3.

Ethane has 1 carbon atom for each of 3 hydrogen atoms.

The molecular formula of ethane is C_2H_6

Each molecule of ethane contains 2 carbon atoms and 6 hydrogen atoms.

Empirical formula represents the simplest, whole-number ratio of atoms of each element in a compound.

Molecular formula represents the actual number of atoms of each element in a molecule of a compound.

KEY POINT

Formula determination

A formula can be calculated from experimental results using the Mole Concept.

Example 1

Find the empirical formula of a compound formed when 6.75 g of aluminium reacts with 26.63 g of chlorine. [A_r: Al, 27.0; Cl, 35.5.]

Find the molar ratio of atoms: Al : Cl

$= \frac{6.75}{27.0} : \frac{26.63}{35.5}$

0.25 : 0.75

Divide by smallest number (0.25): 1 : 3

∴ Empirical formula = $AlCl_3$

Example 2

Find the molecular formula of a compound containing carbon, hydrogen and oxygen only with the composition by mass of carbon: 40.0%; hydrogen: 6.7%. [M_r = 180. A_r: H, 1.00; C, 12.0; O, 16.0.]

Analysis often doesn't give the oxygen content directly but it can easily be calculated.

Percentage of oxygen in compound = 100 – (40.0 + 6.7) = 53.3%

100.0 g of the compound contains 40.0 g C, 6.7 g H and 53.3 g O.

Find the molar ratio of atoms: C : H : O

$= \frac{40.0}{12.0} : \frac{6.7}{1.00} : \frac{53.3}{16.0}$

= 3.33 : 6.7 : 3.33

Divide by smallest number (3.33): 1 : 2 : 1

∴ Empirical formula = CH_2O

Relate to molecular mass

Each CH_2O unit has a relative mass of 12 + (1 × 2) + 16 = 30

M_r, of the compound is 180 containing 180/30 = 6 CH_2O units

∴ molecular formula is $C_6H_{12}O_6$

Progress check

1 Find the empirical formula of the following:
(a) sodium oxide (2.3 g of sodium reacts to form 3.1 g of sodium oxide).
(b) iron oxide (11.16 g of iron reacts to form 15.96 g of iron oxide).

2 Use the following percentage compositions by mass to find the empirical formula of the compound.
(a) A compound of sulfur and oxygen. [S, 40.0%; O, 60.0%]
(b) A compound of iron, sulfur and oxygen. [Fe, 36.8%; S, 21.1%; O, 42.1%]

3 2.8 g of a compound of carbon and hydrogen is formed when 2.4 g of carbon combines with hydrogen [M_r: 56]. What is the empirical and molecular formula of the compound formed?

1 (a) Na_2O; (b) Fe_2O_3.
2 (a) SO_3; (b) $FeSO_4$.
3 CH_2; C_4H_8.

Equations

AQA M1

Chemical reactions involve the **rearrangement** of atoms and ions.

Chemical equations provide two types of information about a reaction:

- qualitative – *which* atoms or ions are rearranging;
- quantitative – *how* many atoms or ions are rearranging.

Balancing equations

The formula of each substance shows the chemicals involved in the reaction.

State symbols can be added to show the physical states of each species under the conditions of the reaction.

State symbols give the physical states of each species in a reaction:
gaseous state, (g);
liquid state, (l);
solid state, (s);
aqueous solution, (aq).

The **qualitative** equation for the reaction of hydrogen with oxygen is shown below:

hydrogen + oxygen ⟶ water

$H_2(g) + O_2(g) \longrightarrow H_2O(l)$

In this equation, hydrogen is balanced, oxygen is **not** balanced.

	$H_2(g)$	+	$O_2(g)$	⟶	$H_2O(l)$	
hydrogen	2				2	✓
oxygen			2		1	✗

The equation must be **balanced** to give the **same number** of particles of each element on each side of the equation.

Balancing an equation often takes several stages.

- Oxygen can be balanced by placing a '2' in front of H_2O.
- However, this unbalances hydrogen:

	$H_2(g)$	+	$O_2(g)$	⟶	**2** $H_2O(l)$	
hydrogen	2				4	✗
oxygen			2		2	✓

- H can be re-balanced by placing a '2' in front of H_2.
- The equation is now balanced.

Notice that no formula has been changed. You balance an equation by adding numbers in front of a formula only.

	2 $H_2(g)$	+	$O_2(g)$	⟶	**2** $H_2O(l)$	
hydrogen	4				4	✓
oxygen			2		2	✓

More balancing numbers.

3 Na_2SO_4 means
Na: 3 × 2 = 6;
S: 3 × 1 = 3;
O: 3 × 4 = 12.

With brackets, the subscript applies to everything in the preceding bracket:

2 $Ca(NO_3)_2$ means
Ca: 2 × 1 = 2;
N: 2 × (1 × 2) = 4;
O: 2 × (3 × 2) = 12.

KEY POINT

In a formula, a subscript applies **only** to the symbol immediately preceding,
e.g. NO_2 comprises **1** N and **2** O.

When balancing an equation, you must **not** change any formula. You can only add a balancing number in front of a formula.

The balancing number multiplies everything in the formula by the balancing number,
e.g. 2$AlCl_3$ comprises Al: 2 × 1 = 2; Cl: 2 × 3 = 6.

Progress check

Balance the following equations:

(a) $Li(s) + O_2(g) \longrightarrow Li_2O(s)$

(b) $CH_4(g) + O_2(g) \longrightarrow CO_2(g) + H_2O(l)$

(c) $Al(s) + HCl(aq) \longrightarrow AlCl_3(aq) + H_2(g)$

(a) $4Li(s) + O_2(g) \longrightarrow 2Li_2O(s)$
(b) $CH_4(g) + 2O_2(g) \longrightarrow CO_2(g) + 2H_2O(l)$
(c) $2Al(s) + 6HCl(aq) \longrightarrow 2AlCl_3(aq) + 3H_2(g)$.

Reacting quantities

AQA M1

The balancing numbers give the ratio of the **amount** of each substance, in mol. Using the example above:

equation	2 $H_2(g)$	+	$O_2(g)$	$\longrightarrow$	2 $H_2O(l)$
moles	2 mol		1 mol	$\longrightarrow$	2 mol

An understanding of molar reacting quantities provides chemists with the required recipes to make quantities of chemicals to order.

Chemists use this **quantitative** information to find:

- the ***reacting quantities*** required to prepare a particular quantity of a product
- the ***quantities of products*** formed by reacting particular quantities of reactants.

By working out the reacting quantities from a balanced chemical equation, quantities can be adjusted to take into account the required scale of preparation.

These examples assume a 100% yield of products. See also percentage yield, page 100.

Example 1

Calculate the masses of nitrogen and oxygen required to form 3 g of nitrogen oxide, NO. [A_r: N, 14; O, 16].

equation	$N_2(g)$	+	$O_2(g)$	$\longrightarrow$	2 NO(g)
moles	1 mol	+	1 mol	$\longrightarrow$	2 mol
reacting masses	(14 x 2) g	+	(16 x 2) g	$\longrightarrow$	2 (14 + 16) g
	28 g	+	32 g	$\longrightarrow$	60 g
for 3 g NO(g), ÷ 20:	1.4 g	+	1.6 g	$\longrightarrow$	3 g

∴ 1.4 g of $N_2(g)$ react with 1.6 g of $O_2(g)$ to form 3 g NO(g).

Example 2

Calculate the volume of oxygen, at r.t.p., formed by the decomposition of an aqueous solution containing 5 g of hydrogen peroxide, H_2O_2. [A_r: H, 1; O, 16.]

You must keep the proportions the same as the reacting quantities.

Any scaling must be applied to all reacting quantities.

equation	2 $H_2O_2(aq)$	$\longrightarrow$	2 $H_2O(l)$	+	$O_2(g)$
moles	2 mol	$\longrightarrow$	2 mol	+	1 mol
reacting quantities	2 [(1 × 2) + (16 × 2)] g	$\longrightarrow$			1 × 24 dm^3
	68 g	$\longrightarrow$			24 dm^3
for 1 g H_2O_2, ÷ 68,	1 g	$\longrightarrow$			$\frac{24}{68}$ dm^3
for 5 g H_2O_2, × 5,	5 g	$\longrightarrow$			$\frac{24}{68}$ × 5 dm^3

∴ 5 g of H_2O_2 decomposes to form 1.8 dm^3 of $O_2(g)$

Progress check

1 For the reaction: $S(s) + O_2(g) \longrightarrow SO_2(g)$
(a) How many moles of S and O_2 react to form 0.5 mole of SO_2?
(b) What mass of SO_2 is formed by reacting 64.2 g of S with O_2?
(c) What volume of SO_2 forms when 0.963 g of S reacts with O_2 at r.t.p.?

2 Balance the equation: $Mg(s) + O_2(g) \longrightarrow MgO(s)$
(a) What mass of MgO forms by burning 8.1 g of Mg?
(b) What volume of O_2 at r.t.p. will react with this mass of Mg?
(c) What masses of Mg and O_2 are needed to prepare 1 g of MgO?

1 (a) 0.5 mol S, 0.5 mol O_2; (b) 96.2 g; (c) 0.72 dm^3.
2 (a) 13.4 g; (b) 4 dm^3; (c) 0.6 g Mg, 0.4 g O_2.

Calculations in acid-base titrations

M1

Calculations for acid-base titrations are best shown with an example:

In a titration, 25.0 cm^3 of 0.100 mol dm^{-3} sodium hydroxide NaOH(aq) were found to react exactly with 20.80 cm^3 of sulfuric acid, $H_2SO_4(aq)$. Find the concentration of the sulfuric acid.

The acid is added from the burette to the alkali.

An indicator is required to show the 'end-point' when all alkali has reacted with the added acid.

To solve the problem, we must use the following pieces of information:

- the balanced equation
$$2NaOH(aq) + H_2SO_4(aq) \longrightarrow Na_2SO_4(aq) + 2H_2O(l)$$
- the concentration c_1 and reacting volume V_1 of NaOH(aq)
- the concentration c_2 and reacting volume V_2 of $H_2SO_4(aq)$.

The concentration and volume of NaOH are known.

From the titration results, the amount of NaOH (in mol) can be calculated:

$$\text{amount of NaOH} = c \times \frac{V}{1000} = 0.100 \times \frac{25.0}{1000} = 0.00250 \text{ mol}$$

Using the equation determine the number of moles of the second reagent

From the equation, the amount of H_2SO_4 (in mol) can be determined:

$$2\,NaOH(aq) + H_2SO_4(aq) \longrightarrow Na_2SO_4(aq) + 2H_2O(l)$$
2 mol   1 mol *(balancing numbers)*

$\therefore$ 0.00250 mol NaOH reacts with 0.00125 mol H_2SO_4

amount of H_2SO_4 that reacted = 0.00125 mol

The concentration (in mol dm^{-3}) of H_2SO_4 can be calculated by scaling to 1000 cm^3:

20.80 cm^3 $H_2SO_4(aq)$ contains 0.00125 mol H_2SO_4

Work out the concentration of $H_2SO_4(aq)$, in mol dm^{-3}.

1 cm^3 $H_2SO_4(aq)$ contains $\frac{0.00125}{20.80}$ mol H_2SO_4

1 dm^3 (1000 cm^3) $H_2SO_4(aq)$ contains $\frac{0.00125}{20.80} \times 1000 = 0.0601$ mol H_2SO_4

$\therefore$ concentration of $H_2SO_4(aq)$ is 0.0601 mol dm^{-3}

Progress check

1 25.0 cm^3 of 0.500 mol dm^{-3} NaOH(aq) reacts with 23.2 cm^3 of $HNO_3(aq)$.
$$HNO_3(aq) + NaOH(aq) \longrightarrow NaNO_3(aq) + H_2O(l)$$
Find the concentration of the nitric acid, HNO_3.

2 25.0 cm^3 of 0.100 mol dm^{-3} KOH(aq) reacts with 26.6 cm^3 of $H_2SO_4(aq)$.
$$H_2SO_4(aq) + 2KOH(aq) \longrightarrow K_2SO_4(aq) + 2H_2O(l)$$
Find the concentration of the sulfuric acid, H_2SO_4.

1 0.539 mol dm^{-3}.
2 0.0470 mol dm^{-3}.

1.7 Redox reactions

After studying this section you should be able to:

- *explain the terms reduction and oxidation in terms of electron transfer*
- *explain the terms oxidising agent and reducing agent*
- *apply the rules for assigning oxidation states*
- *identify changes in oxidation numbers from an equation*
- *construct an overall equation for a redox reaction from half-equations*

LEARNING SUMMARY

Oxidation and Reduction

AQA M2

Redox reactions

Oxidation and *reduction* were originally used for reactions involving oxygen.

> Oxidation is the gain of oxygen.
> Reduction is the loss of oxygen.
>
> KEY POINT

Nowadays, oxidation and reduction have a much broader definition in terms of electron transfer in a *redox* reaction (**red**uction and **ox**idation).

> **Reduction** is the **gain** of electrons.
> **Oxidation** is the **loss** of electrons.
>
> KEY POINT

OIL
RIG

Oxidation
Is
Loss of electrons
Reduction
Is
Gain of electrons

Reduction and oxidation must take place together:

- if one species **gains** electrons
- another species **loses** the same number of electrons

Half-equations

The formation of magnesium chloride from its elements is a redox reaction:

overall reaction $Mg + Cl_2 \longrightarrow MgCl_2$

The overall equation conceals the electron transfer that has taken place. This can be shown by writing half-equations:

electron transfer
$Mg \longrightarrow Mg^{2+} + 2e^-$ oxidation (loss of electrons)
$Cl_2 + 2e^- \longrightarrow 2Cl^-$ reduction (gain of electrons)

Half-equations are useful to identify the species being oxidised or reduced. Notice that the number of electrons lost and gained must balance.

Oxidising and reducing agents

> An **oxidising agent accepts** electrons from another reactant.
> Non-metals are oxidising agents, e.g. F_2, Cl_2, O_2.
>
> A **reducing agent donates** electrons to another reactant.
> Metals are reducing agents, e.g. Na, Fe, Zn.
>
> KEY POINT

Non-metals are oxidising agents.

Metals are reducing agents.

In the example above:

- Mg is the reducing agent – it has *reduced* the Cl_2 to $2Cl^-$ by *adding* electrons
- Cl_2 is the oxidising agent – it has *oxidised* Mg to Mg^{2+} by *removing* electrons.

Using oxidation numbers

AQA M2

Chemists use the concept of **oxidation number** as a means of accounting for electrons.
The oxidation number of a species is assigned by applying a set of rules.

Oxidation number rules

species	*oxidation number*	*examples*
uncombined element	0	C, zero; Na, zero; O_2, zero.
combined oxygen	–2	H_2O; CaO.
combined hydrogen	+1	NH_3, H_2S.
simple ion	charge on ion	Na^+, +1; Mg^{2+}, +2; Cl^-, –1
combined fluorine	–1	NaF, CaF_2.

Exceptions to rules

In compounds with fluorine and in peroxides, the oxidation number of oxygen is not –2 and must be calculated from other oxidation numbers.

In metal hydrides, the oxidation number of hydrogen is –1.

KEY POINT

When applying oxidation numbers to elements, compounds and ions, **the sum of the oxidation numbers must equal the overall charge**.

Examples of applying oxidation numbers

In CO_2, the overall charge is zero.

- There are **2** oxygen atoms, each with an oxidation number of **–2**, giving a total contribution of **–4**.
- The oxidation number of carbon must be **+4** to give the overall charge of zero.

In NO_3^-, the overall charge is 1–.

- There are **3** oxygen atoms, each with an oxidation number of **–2**, giving a total contribution of **–6**.
- The oxidation number of nitrogen must be **+5** to give the overall charge of –1.

Oxidation Number rules can be applied to any compound, whether ionic or covalent.

Oxidation number applies to each atom in a species.

Notice how this has been shown in the examples.

In a compound:	CO_2	
• the total of all the oxidation numbers is zero:	+4	
		–2
		–2
	+4	–4
CO_2: overall charge = 0	0	

In an ion:	NO_3^-	
• the total of all the oxidation numbers is equal to the overall charge on the ion:	+5	
		–2
		–2
		–2
	+5	–6
NO_3^-: overall charge = –1	–1	

Sometimes an element can form compounds or ions in which its atoms can have different oxidation states. The oxidation number is included in the name as a Roman numeral.

For example, there are two nitrate ions:

nitrate(III), NO_2^- N: +3

nitrate(V), NO_3^- N: +5

Using oxidation number with equations

By applying oxidation numbers to an equation:

- the species being oxidised and reduced can be identified
- the number of electrons on both sides of the equation can be checked.

Oxidation numbers can be used to identify redox reactions in which electron loss and electron gain are not easy to see.

	$Cr_2O_3(s)$	+	2 Al (s)	⟶	$Al_2O_3(s)$	+	2 Cr (s)
oxidation numbers	+3 –2 +3 –2 –2		0 0		+3 –2 +3 –2 –2		0 0
sum of oxidation numbers:	0		0		0		0

The sum of the oxidation numbers on both sides of a chemical equation must be the same.

In this reaction, the changes in oxidation number are:

Cr:	$+3 \longrightarrow 0$	reduction	*oxidation number decreases*
Al:	$0 \longrightarrow +3$	oxidation	*oxidation number increases*

> **Oxidation** is an **increase** in oxidation number.
> **Reduction** is a **decrease** in oxidation number.
>
> KEY POINT

Combining half-equations

An overall equation for a redox reaction can be constructed by combining the half-equations showing the transfer of electrons in the reaction.

Example: The reaction of Ag^+ ions react with Zn metal.

The half-equations are:

electron gain: reduction

$$Ag^+(aq) + e^- \longrightarrow Ag(s)$$

electron loss: oxidation

$$Zn(s) \longrightarrow Zn^{2+}(aq) + 2e^-$$

To combine the half-equations in an overall equation, the number of electrons transferred must be the same in each half-equation. This will ensure that every electron lost by Zn(s) is gained by Ag^+.

Balance the electrons: Ag^+ reaction x 2.

- The Ag^+ half-equation is multiplied by '2' to balance the electrons.

$$2Ag^+(aq) + 2e^- \longrightarrow 2Ag(s)$$

- The half-equations are now added:

Cancel the electrons to give the overall equation.

$$2Ag^+(aq) + 2e^- + Zn(s) \longrightarrow 2Ag(s) + Zn^{2+}(aq) + 2e^-$$

- Any species appearing on both sides are cancelled to give the overall equation.

$$2Ag^+(aq) + Zn(s) \longrightarrow 2Ag(s) + Zn^{2+}(aq)$$

- Finally, check that the oxidation numbers balance on either side of the equation:

	$2Ag^+(aq)$	+	$Zn(s)$	$\longrightarrow$	$2Ag(s)$	+	$Zn^{2+}(aq)$
oxidation numbers:	+1		0		0		+2
	+1				0		
oxidation number check:		+2				+2	

Progress check

1 What is the oxidation state of the elements in the following:
(a) Ag^+; (b) F_2; (c) N^{3-}; (d) Fe; (e) MgF_2; (f) $NaClO_3$?

2 Write down the oxidation number of sulfur in the following:
(a) H_2S; (b) SO_2; (c) SO_4^{2-}; (d) SO_3^{2-}; (e) $Na_2S_2O_3$.

3 The following reaction is a redox process:
$Mg + 2HCl \longrightarrow MgCl_2 + H_2$
(a) Identify the changes in oxidation number.
(b) Which species is being oxidised and which is being reduced?
(c) Identify the oxidising agent and the reducing agent.

1 (a) +1; (b) 0; (c) −3; (d) 0; (e) Mg: +2, F: −1; (f) Na: +1, Cl: +5, O: −2.
2 (a) −2; (b) +4; (c) +6; (d) +4; (e) +2.
3 (a) Mg, $0 \longrightarrow +2$, H, $+1 \longrightarrow 0$; (b) Mg oxidised, H reduced;
(c) Oxidising agent: HCl, Reducing agent: Mg.

Sample question and model answer

The determination of reacting quantities using the Mole Concept is one of the most important concepts in chemistry. This can be tested in exam questions almost anywhere and it underpins much of the content of a chemistry course at this level. To succeed at AS Chemistry, it is essential that you grasp this concept.

A student carried out an investigation using a sample of hydrochloric acid, HCl(aq).

(a) The concentration of the hydrochloric acid was first determined by titration of a 25.0 cm^3 sample against 0.148 mol dm^{-3} sodium hydroxide of which 28.40 cm^3 were required.

Calculate the concentration, in mol dm^{-3}, of the hydrochloric acid.

no. of moles of NaOH $= \frac{0.148 \times 28.40}{1000} = 4.20 \times 10^{-3}$ ✓

$NaOH + HCl \longrightarrow NaCl + H_2O$

1 mol NaOH reacts with 1 mol HCl

∴ 4.20×10^{-3} mol NaOH reacts with 4.20×10^{-3} mol HCl ✓

25.0 cm^3 HCl(aq) contains 4.20×10^{-3} mol HCl

1 cm^3 HCl(aq) contains $\frac{4.20 \times 10^{-3}}{25}$ mol HCl

1 dm^3 (1000 cm^3) HCl(aq) contains $\frac{4.20 \times 10^{-3}}{25} \times 1000 = 0.168$ mol HCl.

∴ concentration of HCl(aq) is 0.168 mol dm^{-3} ✓

[3]

(b) The student took a 300 cm^3 sample of the same hydrochloric acid. The student intended to neutralise the hydrochloric acid with calcium hydroxide, $Ca(OH)_2$. The equation is shown below:

$$Ca(OH)_2 + 2HCl \longrightarrow CaCl_2 + 2H_2O$$

Calculate the mass, in g, of calcium hydroxide, $Ca(OH)_2$ required to neutralise this hydrochloric acid.

Notice that the same ideas are used here.

The correct reacting quantities from the equation are essential.

no. of moles of 0.168 mol dm^{-3} HCl in 300 cm^3 $= \frac{0.168 \times 300}{1000} = 0.0504$ ✓

1 mol $Ca(OH)_2$ reacts with 2 mol HCl

∴ $\frac{0.0504}{2}$ mol $Ca(OH)_2$ reacts with 0.0504 mol HCl ✓

Molar mass of $Ca(OH)_2$ = 40 + (16 + 1)2 = 74.1 g mol^{-1} ✓

Mass of $Ca(OH)_2$ $= \frac{0.0504}{2} \times 74.1 = 1.87$ g (to 3 significant figures) ✓

[4]

(c) The student made some calcium hydroxide by heating calcium carbonate and then adding water:

$$CaCO_3 \longrightarrow CaO + CO_2$$
$$CaO + H_2O \longrightarrow Ca(OH)_2$$

Calculate the mass of calcium carbonate that the student would need to produce 10.0 g of calcium hydroxide.

It is important to always show full working. If you do make a mistake early on in a calculation, you will only be penalised once if the method that follows is sound.

In 10.0 g $Ca(OH)_2$, number of moles of $Ca(OH)_2$ $= \frac{10.0}{74.1} = 0.135$ ✓

From the equations, 1 mol $CaCO_3 \longrightarrow$ 1 mol CaO $\longrightarrow$ 1 mol $Ca(OH)_2$

∴ number of moles of $CaCO_3$ = 0.135

∴ Mass of $CaCO_3$ = 0.135 × 100.1 = 13.5 g (to 3 significant figures) ✓

[2]

[Total: 9]

Practice examination questions

1 Chemists use a model of an atom that consists of sub-atomic particles: protons, neutrons and electrons.

(a) Complete the table below to show the properties of these sub-atomic particles.

particle	*relative mass*	*relative charge*
proton		
neutron		
electron		

[3]

(b) The particles in each pair below differ **only** in the number of protons **or** neutrons **or** electrons. Explain what the difference is within each pair.

(i) ^{12}C and ^{13}C

(ii) ^{16}O and $^{16}O^{2-}$

(iii) $^{23}Na^{+}$ and $^{24}Mg^{2+}$. [6]

[Total: 9]

2 (a) In terms of the numbers of sub-atomic particles, state **one** difference and **two** similarities between two isotopes of the same element. [3]

(b) Give the chemical symbol, including its mass number, for an atom which has 5 electrons and 6 neutrons. [1]

(c) An element has an atomic number of 15 and it forms an ion with a charge of 3–.

(i) Deduce the electron configuration of this ion.

(ii) Which block in the Periodic Table is this element in? [2]

(d) Write an equation for the process involved in the first ionisation energy of silicon. [2]

[Total: 8]

3 (a) A student added 0.117 g of lithium to 200 cm^3 of water. The following reaction took place.

$$2Li(s) + 2H_2O(l) \longrightarrow 2LiOH(aq) + H_2(g)$$

(i) Calculate the number of moles of lithium that reacted.

(ii) Calculate the concentration, in mol dm^{-3}, of the 200 cm^3 of aqueous lithium hydroxide solution that formed.

(iii) Calculate the volume of hydrogen gas produced at room temperature and pressure (298 K and 100 kPa).

[6]

(b) In another experiment 25.0 cm^3 of 0.126 mol dm^{-3} lithium hydroxide were neutralised by 21.25 cm^3 of sulfuric acid as shown below.

$$2LiOH(aq) + H_2SO_4(aq) \longrightarrow Li_2SO_4(aq) + 2H_2O(l)$$

Calculate the concentration, in mol dm^{-3}, of the sulfuric acid.

[3]

[Total: 9]

Practice examination questions

4 A student investigated different methods to neutralise 0.580 mol dm^{-3} sulfuric acid, H_2SO_4.

(a) The student decided to add 1.25 mol dm^{-3} NaOH to a 25.0 cm^3 sample of the 0.580 mol dm^{-3} sulfuric acid. The reaction that takes place is shown below.

$$2NaOH(aq) + H_2SO_4(aq) \longrightarrow Na_2SO_4(aq) + 2H_2O(l)$$

(i) Calculate the amount, in moles, of H_2SO_4 in 25.0 cm^3 of 0.580 mol dm^{-3} of sulfuric acid.

(ii) What is the amount, in moles, of NaOH that is needed to neutralise this amount of acid?

(iii) Calculate the volume, in cm^3, of 1.25 mol dm^{-3} NaOH that would be needed to neutralise this amount of acid.

[3]

(b) The student then decided to add sodium carbonate, Na_2CO_3, to a second 25.0 cm^3 sample of the 0.580 mol dm^{-3} sulfuric acid. The equation for this reaction is shown below.

$$Na_2CO_3(s) + H_2SO_4(aq) \longrightarrow Na_2SO_4(s) + H_2O(l) + CO_2(g)$$

Calculate the mass of Na_2CO_3 that would be needed to neutralise this amount of sulfuric acid.

[2]

(c) In a third experiment the student added magnesium metal to the sulfuric acid. The equation for the reaction that takes place is shown below.

$$Mg(s) + H_2SO_4(aq) \longrightarrow MgSO_4(aq) + H_2(g)$$

Use oxidation numbers to show that this is a redox reaction.

[2]
[Total: 7]

5 Sodium azide, NaN_3, decomposes when heated to release nitrogen gas.

$$2NaN_3(s) \longrightarrow 2Na(l) + 3N_2(g)$$

(a) A student prepared 1.50 dm^3 of nitrogen in the laboratory by this method. This gas volume was measured at room temperature and pressure (r.t.p.).

(i) How many moles of N_2 did the student prepare?
[Assume that 1 mole of gas molecules occupies 24.0 dm^3 at r.t.p.]

(ii) What mass of NaN_3 did the student heat? [4]

(b) In this reaction, 0.96 g of sodium metal was also formed. The student carefully reacted the sodium with water to form 50.0 cm^3 of aqueous sodium hydroxide:

$$2Na(s) + 2H_2O(l) \longrightarrow 2NaOH(aq) + H_2(g)$$

(i) Calculate the concentration, in mol dm^{-3}, of the aqueous sodium hydroxide.

(ii) Use oxidation numbers to show that this is a redox reaction. [4]
[Total: 8]

Chapter 2

Bonding, structure and the Periodic Table

The following topics are covered in this chapter:

- *Chemical bonding*
- *Shapes of molecules*
- *Electronegativity, polarity and polarisation*
- *Intermolecular forces*
- *Bonding, structure and properties*
- *The modern Periodic Table*
- *The s-block elements: Group 1 and Group 2*
- *The Group 7 elements and their compounds*
- *Extraction of metals*
- *Qualitative analysis*

2.1 Chemical bonding

After studying this section you should be able to:

- *appreciate that compounds often contain atoms or ions with electron structures of a noble gas*
- *understand the nature of an ionic bond in terms of electrostatic attraction between ions*
- *determine the formula of an ionic compound from the ionic charges*
- *understand the nature of a covalent bond and a dative covalent (coordinate) bond in terms of the sharing of an electron pair*
- *construct 'dot-and-cross' diagrams to demonstrate ionic and covalent bonding*
- *describe metallic bonding*

LEARNING SUMMARY

Why do atoms bond together?

AQA M1

The Octet Rule

Under normal conditions, the only elements that can exist as single atoms are the Noble Gases in Group 0 of the Periodic Table. The atoms of all other elements are bonded together. To find out why, we need to look at the electron configurations of the atoms of the Noble Gases.

Atoms of the Noble Gases are stable because all their electrons are paired.

The Noble Gases get their name from their unreactivity.

In the atoms of a noble gas:

- all electrons are paired,
- the bonding shells are full.

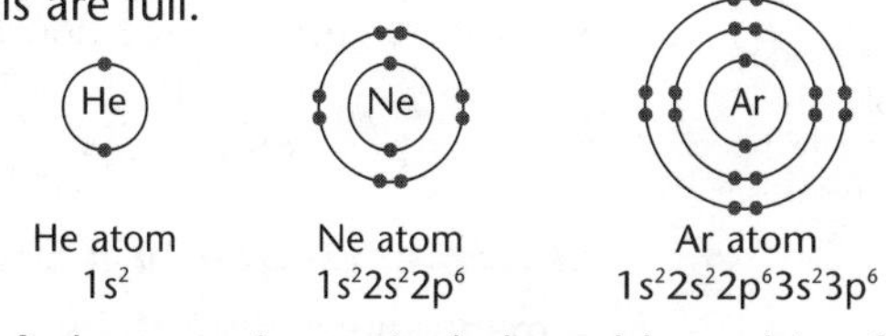

He atom $1s^2$ Ne atom $1s^2 2s^2 2p^6$ Ar atom $1s^2 2s^2 2p^6 3s^2 3p^6$

This arrangement of electrons is particularly stable and is often referred to as a stable 'octet' (as in neon and argon).

This tendency to acquire a noble gas electron structure is often referred to as the 'Octet Rule'.

Apart from the Noble Gases, atoms are bonded together. Thus the elements oxygen and nitrogen exists as O_2 and N_2. Compounds exist when atoms of different elements bond together as in CO_2, CO and NaCl. In all these examples, electrons have been shared, transferred or rearranged in such a way that the atoms of the elements have acquired a noble gas electron structure. If this creates an outer shell of 8 electrons (as in all these cases), the '*Octet Rule*' is obeyed. All these themes are discussed in the sections that follow.

Types of bonding

Many mistakes are made in chemistry by confusing the type of bonding.

Covalent and ionic compounds are bonded differently and they behave differently. You cannot apply 'ionic bonding ideas' to a compound that is covalent.

Chemical bonds are classified into three main types: ionic, covalent and metallic.

- **Ionic** bonding occurs between the atoms of a **metal** and a **non-metal**. For example, NaCl, MgO, Fe_2O_3.
- **Covalent** bonding occurs between the atoms of **non-metals**. For example, O_2, H_2, H_2O, diamond and graphite.
- **Metallic** bonding occurs between the atoms of **metals**. For example, all metals such as iron, zinc, aluminium, etc.; alloys as in brass (copper and zinc), bronze (copper and tin).

> **KEY POINT**
> Always decide the type of bonding before:
> - drawing a 'dot-and-cross' diagram
> - predicting how a compound reacts
> - predicting the properties of a compound.

Ionic bonding

AQA M1

Ionic bonds

An ionic bond is formed when electrons are **transferred** from a metal atom to a non-metal atom forming **oppositely-charged ions**.

Ionic bonds are present in a compound of a metal and a non-metal.

> **KEY POINT**
> An ionic bond is the electrical attraction between oppositely charged ions.

A positive ion is called a **cation** because it is attracted to a **cathode** (a negative electrode). A negative ion is called an **anion** because it is attracted to an **anode** (a positive electrode).

For showing chemical bonding, 'dot-and-cross' diagrams often only show the outer shell.

Example: sodium chloride, NaCl

Sodium chloride, NaCl, is formed by the transfer of one electron from a sodium atom ($[Ne]3s^1$) to a chlorine atom ($[Ne]3s^23p^5$) with the formation of ions:

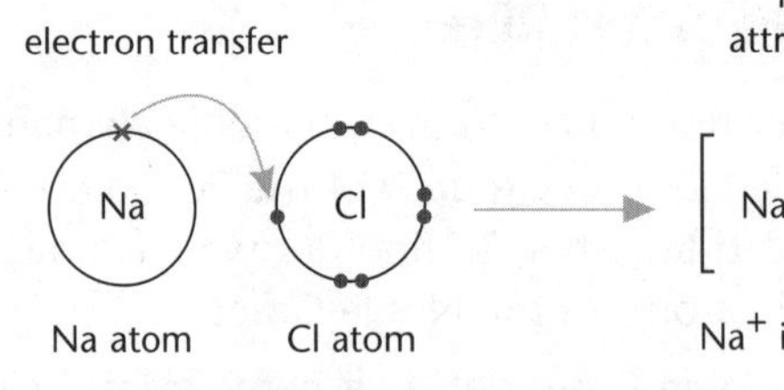

ionic bond forms: attraction between ions

Na⁺ ion Cl⁻ ion

A sodium atom has lost an electron to acquire the electron structure of neon:

$$Na \longrightarrow Na^+ + e^-$$
$$[Ne]3s^1 \longrightarrow [Ne] + e^-$$

A chlorine atom has gained an electron to acquire the electron structure of argon:

The ions that are formed have stable noble gas electron structures with full outer electron shells.

$$Cl + e^- \longrightarrow Cl^-$$
$$[Ne]3s^23p^5 + e^- \longrightarrow [Ne]3s^23p^6 \text{ or } [Ar]$$

Example: magnesium chloride, $MgCl_2$

A magnesium atom ($[Ne]3s^2$) has two electrons in its outer shell. One electron is transferred to each of two chlorine atoms ($[Ne]3s^23p^5$) forming ions.

The Mg^{2+} and Cl^- ions have noble gas electron configurations.

transfer of 2 electrons

ionic bond forms: attraction between ions

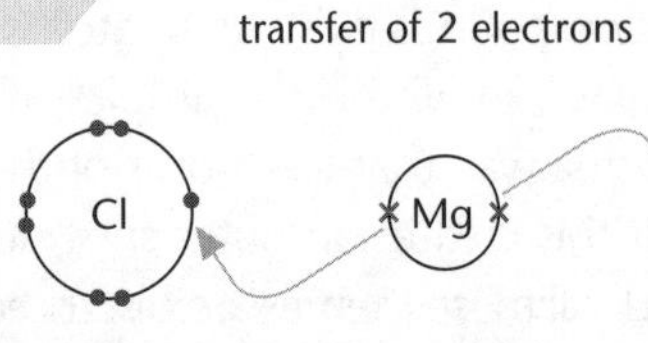

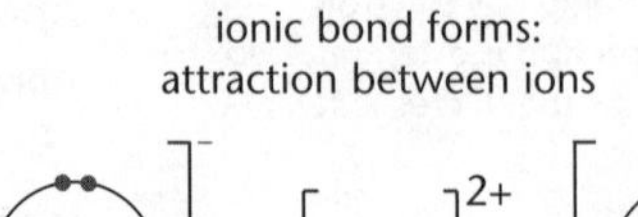

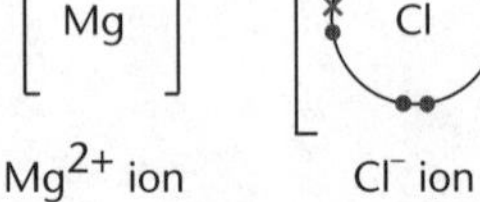

Cl atom 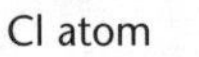 Mg atom 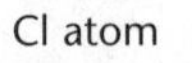 Cl atom

Cl⁻ ion Mg²⁺ ion Cl⁻ ion

Giant ionic lattices

Each ion is surrounded by oppositely-charged ions, forming a giant ionic lattice.

- Although it is convenient to look at ionic bonding between two ions only, each ion is able to attract oppositely charged ions in all directions.
- This results in a **giant ionic lattice** structure comprising hundreds of thousands of ions (depending upon the size of the crystal).
- This arrangement is characteristic of all ionic compounds.

Part of the sodium chloride lattice

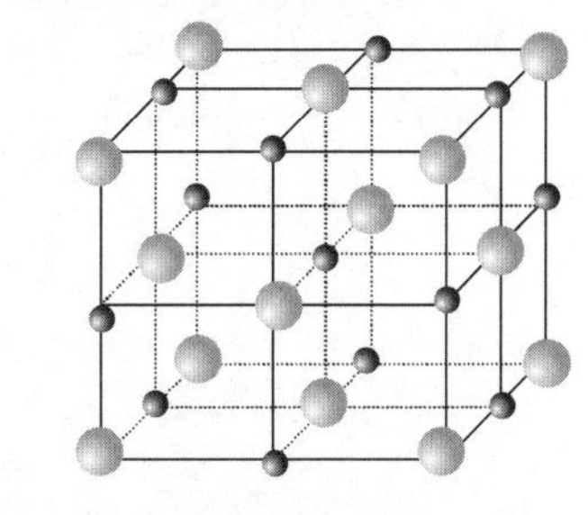

- Each Na^+ ion surrounds 6 Cl^- ions
- Each Cl^- ion surrounds 6 Na^+ ions

Ionic charges and the Periodic Table

An ionic charge can be predicted from an element's position in the Periodic Table.

The diagram below shows elements in Periods 2 and 3 of the Periodic Table.

group	*1*	*2*	*3*	*4*	*5*	*6*	*7*	*0*
number of outer shell electrons	1	2	3	4	5	6	7	8
element	Li	Be	B	C	N	O	F	Ne
ion	Li^+				N^{3-}	O^{2-}	F^-	
element	Na	Mg	Al	Si	P	S	Cl	Ar
ion	Na^+	Mg^{2+}	Al^{3+}		P^{3-}	S^{2-}	Cl^-	

Remember that any metal ions will be positive and non-metal ions negative.

- Metals in Groups 1–3 **lose** sufficient electrons from their atoms to form ions with the electron configuration of the **previous** noble gas;
- Non-metals in Groups 5–7 **gain** sufficient electrons to their atoms to form ions with the electron configuration of the **next** noble gas.
- Be, B, C and Si do not normally form ionic compounds because of the large amount of energy required to transfer electrons forming their ions.

Groups of covalent-bonded atoms can also lose or gain electrons to give ions such as those shown below.

Learn these

1+		*1–*		*2–*		*3–*	
ammonium	NH_4^+	hydroxide	OH^-	carbonate	CO_3^{2-}	phosphate	PO_4^{3-}
		nitrate	NO_3^-	sulfate	SO_4^{2-}		
		nitrite	NO_2^-	sulfite	SO_3^{2-}		
		hydrogen-carbonate	HCO_3^-				

Predicting ionic formulae

Although an ionic compound comprises charged ions, its overall charge is zero.

KEY POINT

In an ionic compound,

total number of positive charges from positive ions = total number of negative charges from negative ions.

Working out an ionic formula from ionic charges

calcium chloride:	*ion*	*charge*	*aluminium sulfate:*	*ion*	*charge*
equalise charges:	Ca^{2+}	2+	equalise charges:	Al^{3+}	3+
	Cl^-	1–		Al^{3+}	3+
	Cl^-	1–		SO_4^{2-}	2–
				SO_4^{2-}	2–
				SO_4^{2-}	2–
total charge must be zero:		0	total charge must be zero:		0
formula:	$CaCl_2$		formula:	$Al_2(SO_4)_3$	

Covalent bonding

AQA M1

Covalent bonds

A covalent compound comprises molecules: groups of atoms held together by covalent bonds.

KEY POINT: A molecule is the smallest part of a covalent compound that can take part in a chemical reaction.

A covalent bond is formed when electrons are **shared** rather than transferred.

Covalent bonds are present in a compound of two non-metals.

KEY POINT: A covalent bond is a shared pair of electrons.

Example: hydrogen, H_2

The covalent bond in a hydrogen molecule H_2 forms when two hydrogen atoms ($1s^1$) bond together, each contributing 1 electron to the bond:

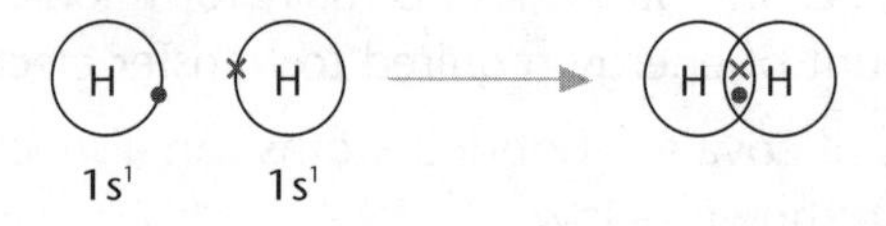

A single covalent bond is written as A–B.

In forming the covalent bond, each hydrogen atom in the H_2 molecule has:

- control of 2 electrons – one of its own and the second from the other atom
- the electron configuration of the noble gas helium.

When a covalent bond forms, each atom *often* acquires a stable noble gas electron configuration with a full outer bonding shell.

This diagram shows how two atoms are bonded together by a covalent bond.

Unlike an ionic bond, a covalent bond is *directional* and acts solely between the two atoms involved in the bond. Thus hydrogen exists simply as H_2 molecules.

The diagrams below show examples of covalent bonding in some simple molecules.

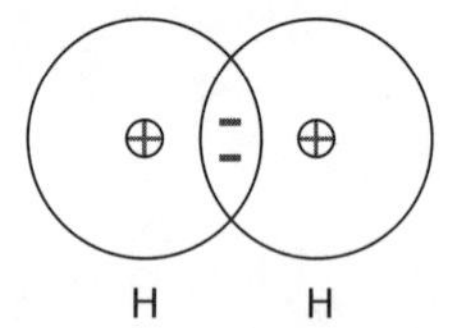

In a hydrogen molecule, the nucleus of each hydrogen atom is attracted towards the electron pair of the covalent bond.

Notice that each atom contributes one electron to the covalent bond.

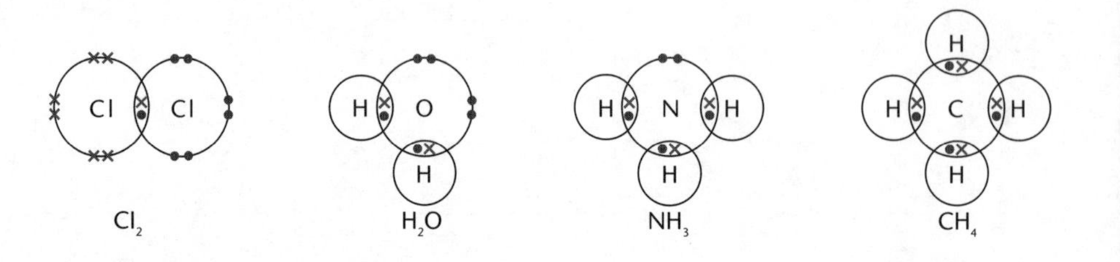

Multiple covalent bonds

Other common examples include:

C_2H_4: $H_2C{=}CH_2$
CO_2: O=C=O
C_2H_2: H–C≡C–H

A covalent bond in which **one** pair of electrons is shared is known as a **single bond**, e.g. Cl_2, written as Cl–Cl.

Atoms can also share more than one pair of electrons to form a **multiple bond**.

- Sharing of **two** pairs of electrons forms a **double bond**, e.g. O_2, written as O=O.
- Sharing of **three** pairs of electrons forms a **triple bond**, e.g. N_2, written as N≡N.

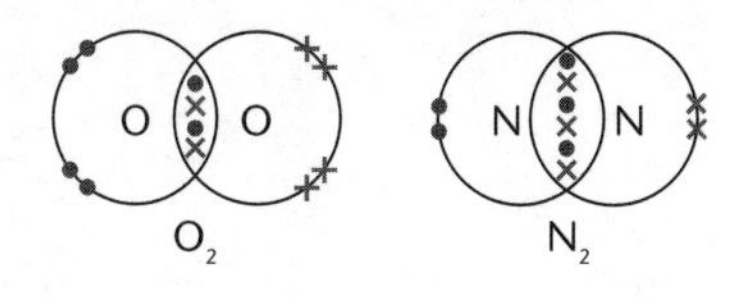

Dative covalent or coordinate bonds

A dative covalent bond forms when the shared pair of electrons comes from just **one** of the bonded atoms.

A dative covalent or coordinate bond is one in which **one** of the atoms supplies **both** the shared electrons to the covalent bond. A dative covalent bond is written as A→B, where the direction of the arrow shows the direction in which the electron pair is donated.

Example: ammonium ion, NH_4^+

- The ammonium ion, NH_4^+, forms when ammonia NH_3 bonds with a proton H^+, with both the bonding electrons coming from NH_3.
- Note that the ammonium ion has three covalent bonds and one dative covalent bond:

The involvement of the electron pair as a dative covalent bond is often shown as an arrow:

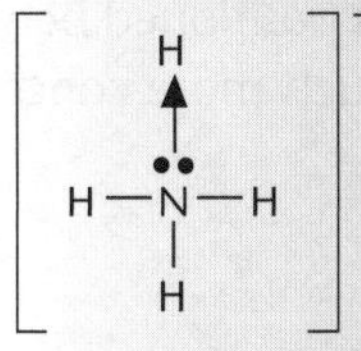

Note that once NH_4^+ has formed, all 4 bonds are equivalent and you cannot tell which was formed by a dative covalent bond.

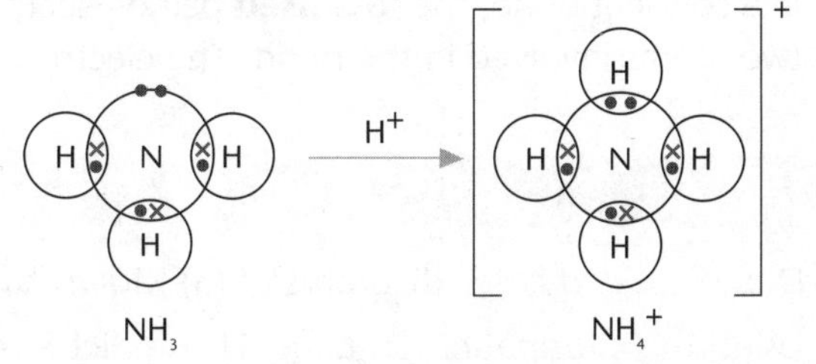

In exams, these are easy marks provided that the facts have been learnt.

The Octet Rule doesn't always work

The Octet Rule is useful in chemistry but it can be broken. It is probably more important that all unpaired electrons are paired.

When covalent bonds form, unpaired electrons **often** pair up to form a noble gas electron configuration according to the Octet Rule (see page 39). Many molecules, however, form electron configurations that do not form an octet. Some molecules have 6, 10, 12 or even 14 electrons in the outer bonding shell. These compounds form by the pairing of unpaired electrons.

Phosphorus forms two chlorides, PCl_3 and PCl_5.

PCl_3 obeys the Octet Rule.

In PCl_5,

- the outer shell of phosphorus has expanded allowing all 5 electrons to be used in bonding
- the Octet Rule is broken and phosphorus now has 10 electrons in its outer shell:

In PCl_3, phosphorus is using 3 valence electrons.

In PCl_5, phosphorus is using 5 valence electrons (the same as the Group number of phosphorus).

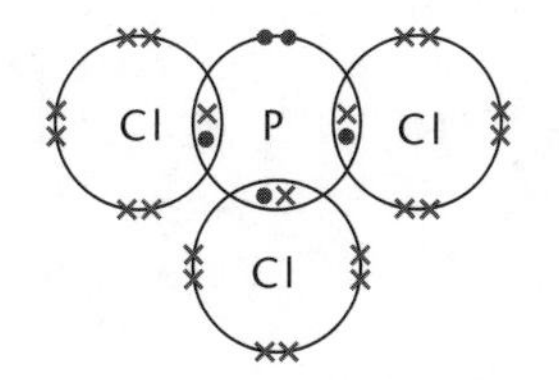

PCl_3: only 3 electrons from the outer bonding shell of phosphorus are used in bonding.

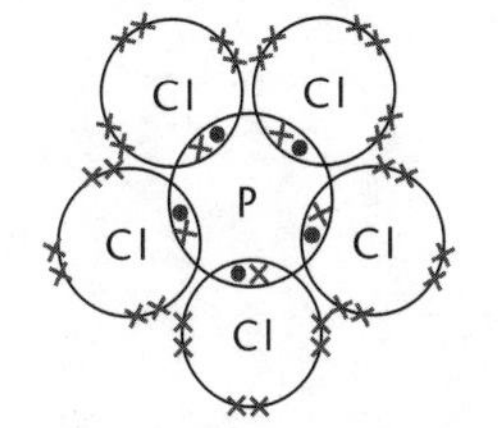

PCl_5: only 5 electrons from the outer bonding shell of phosphorus are used in covalent bonding.

An electron that takes part in forming chemical bonds is called a **valence electron**.

For elements in the s-block and p-block of the Periodic Table, the maximum number of valence electrons available is usually equal to the group number.

Metallic bonding

AQA M1

A metallic bond holds atoms together in a solid metal or alloy.

In solid metals, the atoms are ionised.

- The positive ions occupy fixed positions in a lattice.
- The outer shell electrons are **delocalised** – they are spread throughout the metallic structure and are able to move freely throughout the lattice.

KEY POINT

A **metallic bond** is the electrostatic attraction between the positive metal ions and delocalised electrons.

A metallic lattice is held together by electrostatic attraction between positive metal ions and electrons.

In the metallic lattice, each metal atom exists as + ions by releasing its outer valence electrons to the delocalised pool of electrons (often called a 'sea of electrons').

The 'sea of electrons'

Delocalised and localised electrons

The **delocalised** electrons in metals are spread throughout the metal structure. The delocalisation is such that electrons are able to move throughout the structure and it is impossible to assign any electron to a particular positive ion.

In a covalent bond, the **localised** pair of electrons is always positioned between the two atoms involved in the bond. The electron charge is much more concentrated.

Progress check

1 Draw *'dot-and-cross'* diagrams of (a) MgO; (b) Na_2O.

2 Using the information on page 41, predict formulae for the following ionic compounds:

(a) lithium chloride
(b) potassium sulfide
(c) lithium nitride
(d) aluminium oxide
(e) sodium carbonate
(f) calcium hydroxide
(g) aluminium sulfate
(h) ammonium phosphate.

3 Draw *'dot-and-cross'* diagrams for
(a) C_2H_6; (b) HCN; (c) H_3O^+.

2 (a) LiCl; (b) K_2S; (c) Li_3N; (d) Al_2O_3; (e) Na_2CO_3; (f) $Ca(OH)_2$; (g) $Al_2(SO_4)_3$; (h) $(NH_4)_3PO_4$.

2.2 Shapes of molecules

After studying this section you should be able to:

- *use electron-pair repulsion to explain the shapes and bond angles of simple molecules*
- *understand that lone pairs have a larger repulsive effect than bonded pairs, leading to distortion of the molecular shape*
- *understand how multiple bonds affect the shape of a molecule*

LEARNING SUMMARY

Electron-pair repulsion theory

AQA M1

Electron-pair repulsion theory states that:

- the shape of a molecule depends upon the number of electron pairs surrounding the central atom
- electron pairs repel one another and move as far apart as possible.

Bonds and lone pairs are negative charge clouds that repel one another.

The shape of any molecule can be predicted by applying this theory.

Molecules with bonded pairs

Learn these shapes and bond-angles.

You can predict the shape of a molecule from its **'dot-and-cross'** diagram. Try to relate the number of electron pairs in the examples below with the three-dimensional shape of each molecule.

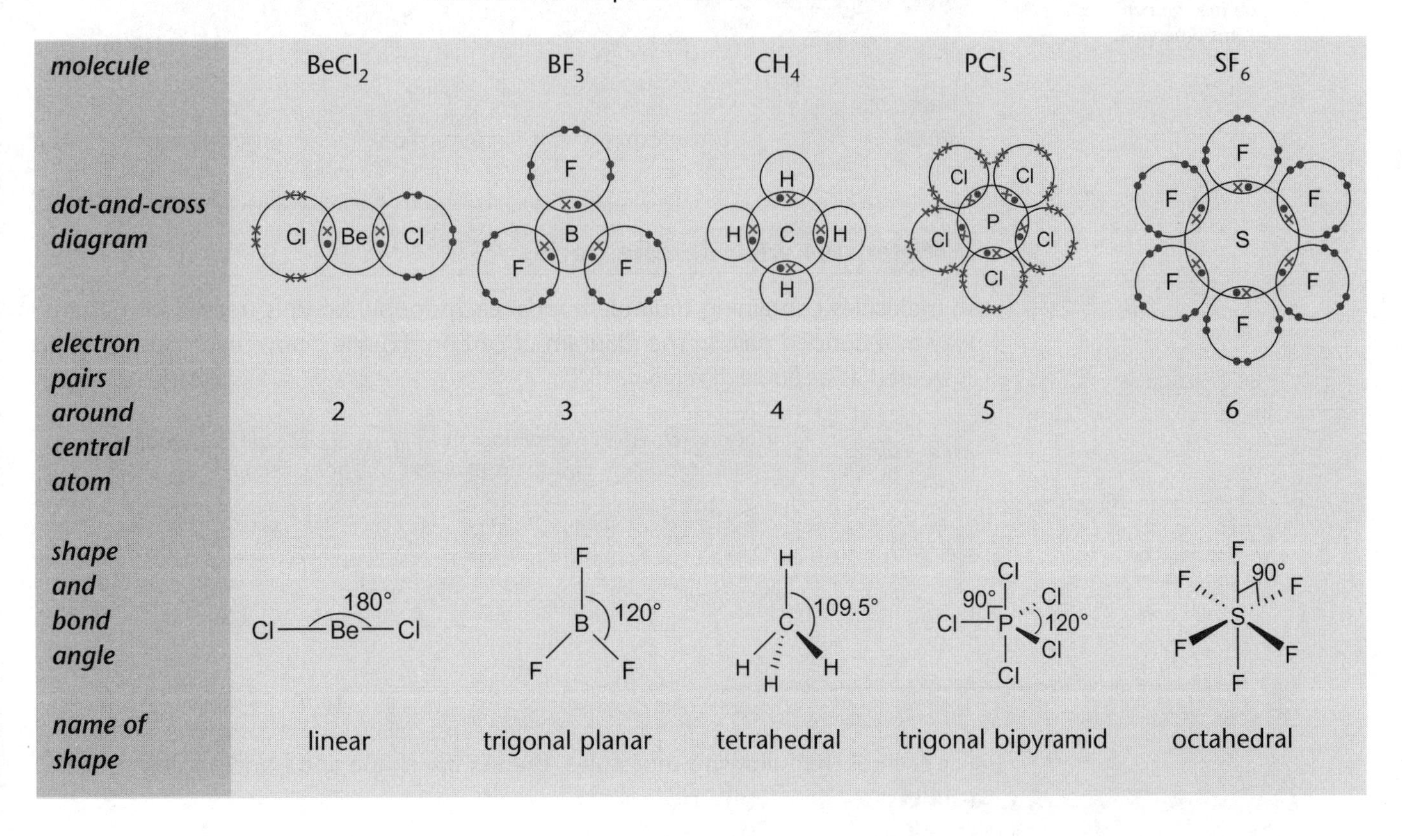

molecule	$BeCl_2$	BF_3	CH_4	PCl_5	SF_6
dot-and-cross diagram					
electron pairs around central atom	2	3	4	5	6
shape and bond angle	180°	120°	109.5°	90°, 120°	90°
name of shape	linear	trigonal planar	tetrahedral	trigonal bipyramid	octahedral

Molecules with lone pairs

Each of the examples below are molecules with four electron pairs surrounding the central atom. The molecular shape will therefore be based upon a tetrahedron. However, a lone pair is closer to an atom than a bonded pair of electrons and has a larger repulsive effect than a bonded pair.

> **KEY POINT**
>
> The relative magnitudes of electron-pair repulsions are:
> **lone**-pair/**lone**-pair > **bonded**-pair/**lone**-pair > **bonded**-pair/**bonded**-pair.

Lone pairs distort the shape of a molecule and reduce the bond angle. Look at the examples below and notice how each lone pair decreases the bond-angle by 2.5°.

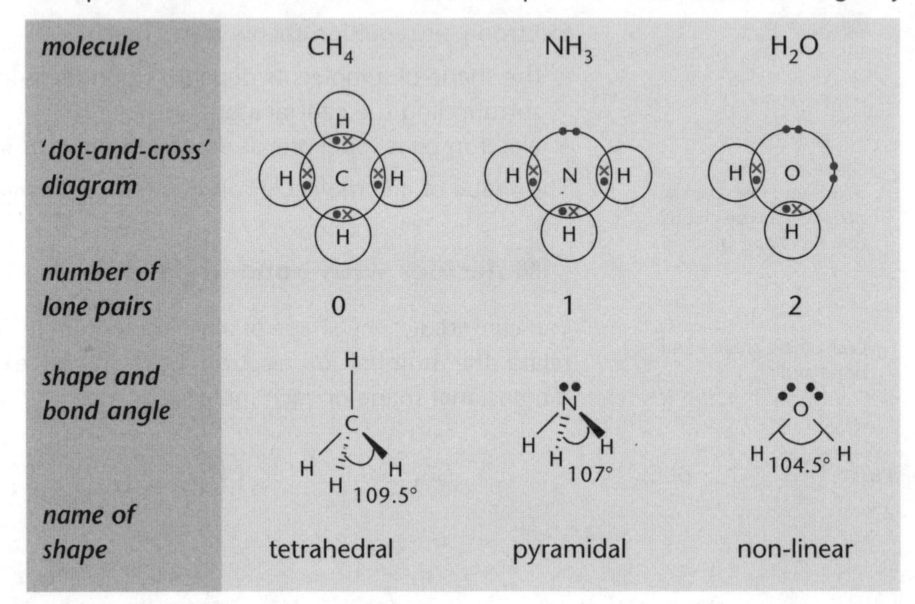

Always draw 3-D shapes – you may be penalised in exams otherwise.

Molecules with double bonds

In molecules containing multiple bonds, each double bond is treated in the same way as a bonded pair. In the diagram of carbon dioxide below, each double bond is treated as a *'bonding region'*.

molecule	*'dot-and-cross' diagram*	*number of bonding regions*	*shape and bond angle*	*name of shape*
CO_2	O C O	2	O=C=O, 180°	linear

Progress check

1 For each of the following molecules, predict the shape and bond angles.

(a) BeF_2
(b) $AlCl_3$
(c) SiH_4
(d) H_2S
(e) PH_3
(f) CS_2
(g) SO_3
(h) SO_2

1 (a) linear, 180°
(b) trigonal planar, 120°
(c) tetrahedral, 109.5°
(d) non-linear, 104.5°
(e) pyramidal, 107°
(f) linear, 180°
(g) trigonal planar, 120°
(h) non-linear, 120°

2.3 Electronegativity, polarity and polarisation

LEARNING SUMMARY

After studying this section you should be able to:

- *appreciate that many chemical bonds have bonding intermediate between ionic and covalent bonding*
- *describe electronegativity in terms of attraction for bonding electrons*
- *understand the nature of polarity in molecules of covalent compounds*

Ionic or covalent?

AQA M1

Between the extremes of ionic and covalent bonding, there is a whole range of intermediate bonds, which have both ionic and covalent contributions.

An ionic bond with 100% ionic character would require the complete transfer of an electron from a metal atom to a non-metal atom. In practice, this never completely happens, although some compounds have close to 100% ionic character.

In a hydrogen molecule, H_2, the two hydrogen atoms are identical and form a covalent bond with 100% covalent character by equally sharing the bonded pair of electrons. However, in molecules such as HCl, the bonded pair of electrons is not shared equally.

An ionic bond is often formed with some covalent character:

- the electron transfer is incomplete
- there is a degree of electron sharing.

A covalent bond is often formed with some ionic character:

- the electrons are not equally shared
- there is a degree of electron transfer.

The sections that follow discuss intermediate bonding in covalent compounds.

Electronegativity

AQA M1

The nuclei of the atoms in a molecule attract the electron pair in a covalent bond.

KEY POINT

Electronegativity is a measure of the attraction of an atom in a molecule for the pair of electrons in a covalent bond.

The most electronegative atoms attract bonding electrons most strongly.

- In general, small atoms are electronegative atoms.
- The most electronegative atoms are those of highly reactive non-metallic elements (such as O, F and Cl).
- Reactive metals (such as Na and K) have the least electronegative atoms.

How is electronegativity measured?

Notice that the Noble Gases are not included. Although neon and helium atoms are smaller than those of fluorine, they do not form bonds and so they have no affinity for bonded electrons.

See how the Pauling values of electronegativity relate to the element's position in the Periodic Table.

The 'Pauling scale' is often used to compare the electronegativities of different elements. The diagram below shows how the electronegativity of an element relates to its position in the Periodic Table. The numbers give the Pauling electronegativity values.

electronegativity increases

Li	Be	B	C	N	O	F
1.0	1.5	2.0	2.5	3.0	3.5	4.0
Na						Cl
0.9						3.0
K						Br
0.8						2.8

Fluorine, the most electronegative element, has small atoms with a greater attraction for the pair of electrons in a covalent bond than larger atoms.

> **KEY POINT**
>
> The greater the **difference** between electronegativities, the greater the **ionic** character of the bond.
>
> The greater the **similarity** in electronegativities, the greater the **covalent** character of the bond.

Polar and non-polar molecules

AQA M1

Non-polar bonds

When bonding atoms are the same, the attraction for the bonded pair of electrons is the same and the bond is non-polar.

Hydrocarbons are non-polar because C and H have very similar electronegativities.

A covalent bond is non-polar when:

- the bonded electrons are shared **equally** between both atoms
- the bonded atoms have similar electronegativities.

A covalent bond **must** be non-polar if the bonded atoms are the same, as in molecules of H_2 and Cl_2 shown here:

H H

H_2 molecule

Cl Cl

Cl_2 molecule

Polar bonds

In a polar covalent bond, the bonded electrons are attracted towards the more electronegative of the atoms.

The electronegativities of hydrogen halides are discussed in more detail on page 63.

A covalent bond is polar when:

- the electrons in the bond are shared **unequally** making a *polar bond*
- the bonded atoms are different, each has a different electronegativity.

In hydrogen chloride, HCl, the Cl atom is more **electronegative** than the H atom, making the H–Cl covalent bond polar.

The molecule is *polarised* with a small positive charge δ+ on the hydrogen atom and a small negative charge δ– on the chlorine atom.

δ+ δ-
H Cl

bonded pair of electrons attracted closer to chlorine

The hydrogen chloride molecule is *polar* with a *permanent dipole*.

Symmetrical and unsymmetrical molecules

If the molecule is symmetrical, any dipoles will cancel and the molecule will not have a permanent dipole. The diagram below shows why the symmetrical molecule, CCl_4, is non-polar.

- Each C–Cl bond is polar but the dipoles act in different directions.
- The overall effect is for the dipoles to cancel each other.

∴ CCl_4 is a non-polar molecule.

Dipoles in symmetrical molecules cancel.

CCl_4 is non-polar

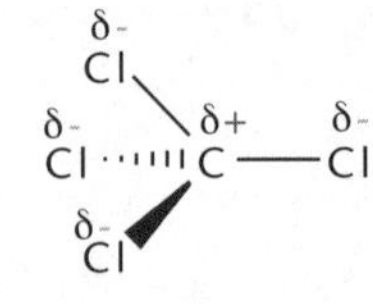

Although each C–Cl bond is polar, the dipoles cancel

Progress check

1 Decide whether each molecule is polar and state which atom, if any, has the δ– charge.
(a) Br_2; (b) H_2O; (c) O_2; (d) HBr; (e) NH_3.

2 (a) Predict the shape of a molecule of BF_3 and of PF_3.
(b) Explain why BF_3 is non-polar whereas PF_3 is polar.

1 (a) non-polar; (b) polar, $O^{\delta-}$; (c) non-polar; (d) polar, $Br^{\delta-}$; (e) polar, $N^{\delta-}$.
2 (a) BF_3: trigonal planar; PF_3: pyramidal.
(b) BF_3 has a symmetrical shape. Although each bond is polar, the dipoles cancel. PF_3 is not symmetrical. The F atoms are all on the same side of the molecule resulting in a permanent dipole.

2.4 Intermolecular forces

After studying this section you should be able to:

LEARNING SUMMARY

- *understand the nature of van der Waals' forces (induced dipole-dipole interactions)*
- *understand the nature of permanent dipole-dipole interactions and hydrogen bonds*
- *describe the anomalous properties of water arising from hydrogen bonding.*

van der Waals' forces or induced dipole-dipole forces

AQA M1

Van der Waals' forces (induced dipole-dipole forces) exist between all molecules whether polar or non-polar.

What causes van der Waals' forces?

Induced dipole-dipole interactions are often referred to as van der Waals' forces.

Van der Waals' forces are:

- weak intermolecular forces
- caused by attractions between very small dipoles in molecules.

Ionic and covalent bonds are of comparable strength.

Intermolecular forces are far weaker.

The diagram below shows how these forces arise between single atoms in a noble gas.

Movement of electrons produces an oscillating dipole

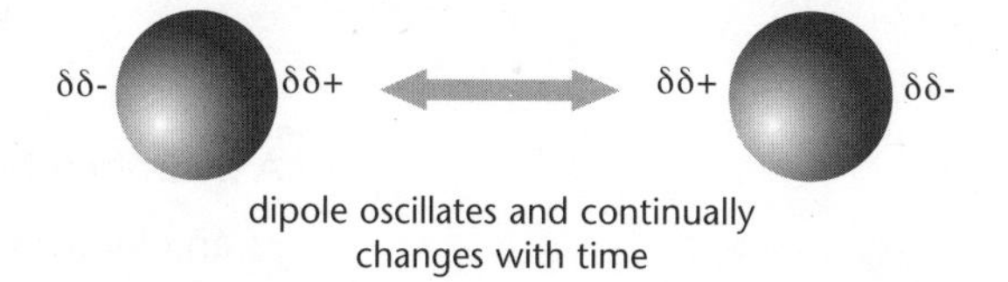

dipole oscillates and continually changes with time

An oscillating or instantaneous dipole is caused by an uneven distribution of electrons at an instant of time.

Oscillating dipole induces a dipole in a neighbouring molecule which is induced onto further molecules

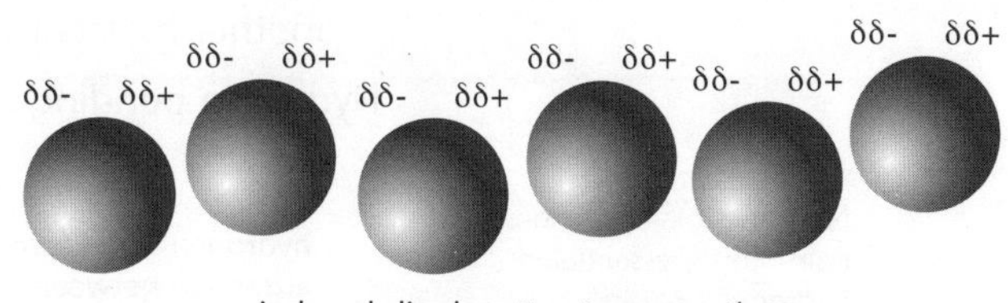

induced dipoles attract one another

What affects the strength of van der Waals' forces?

Van der Waals' forces result from interactions of electrons between molecules.

Van der Waals' forces increase in strength with increasing number of electrons.

The greater the number of electrons in each molecule:

- the larger the oscillating and induced dipoles
- the greater the attractive forces between molecules
- the greater the van der Waals' forces.

This can be seen by comparing the boiling points of the Noble Gases.

Noble gas	*b. pt./°C*	*number of electrons*	*trend*
He	–269	2	• easier to distort electron clouds
Ne	–246	10	
Ar	–186	18	• induced dipoles increase
Kr	–152	36	
Xe	–107	54	• boiling point increases
Rn	– 62	86	

Permanent dipole-dipole interactions

AQA M1

The small δ+ and δ− charges on a polar molecule attract oppositely charged dipoles on another polar molecule.

This gives a weak *intermolecular force* called a permanent dipole-dipole interaction.

Permanent dipole-dipole interactions are 'non-directional'.

They are simply weak attractions between dipole charges on different molecules.

Example: Intermolecular forces between HCl molecules

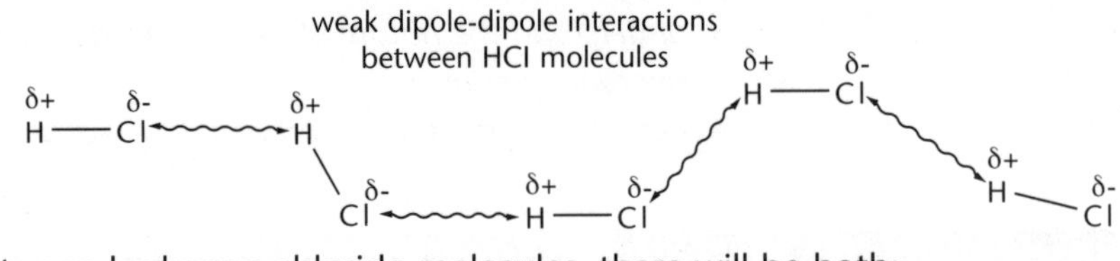

Between hydrogen chloride molecules, there will be both:

- van der Waals' forces and
- permanent dipole-dipole interactions.

Although both are weak forces, the permanent dipole-dipole interactions are still stronger than the van der Waals' forces.

Hydrogen bonds

AQA M1

A hydrogen bond is a special type of permanent dipole-dipole interaction found between molecules containing the following groups:

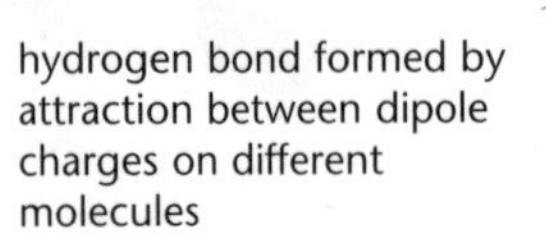

A hydrogen bond is a comparatively strong intermolecular attraction between:

- an electron deficient hydrogen atom, $H^{\delta+}$, on one molecule and
- a lone pair of electrons on a highly electronegative atom of F, O or N on another molecule.

Hydrogen bonding occurs between molecules such as H_2O:

Note the role of the lone pair – this is essential in hydrogen bonding.

A hydrogen bond is shown between molecules as a dashed line.

hydrogen bond formed by attraction between dipole charges on different molecules

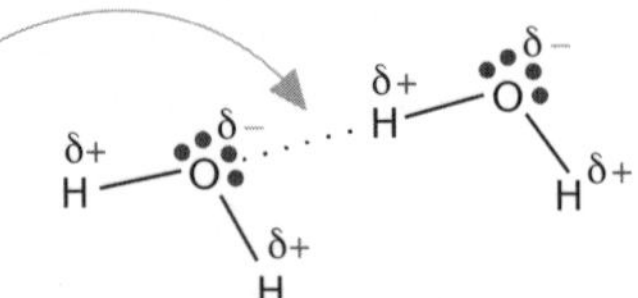

Hydrogen bonding is especially important in organic compounds containing –OH or –NH bonds: e.g. alcohols, carboxylic acids, amines, amino acids.

Special properties of water arising from hydrogen bonding

A hydrogen bond has only one-tenth the strength of a covalent bond. However, hydrogen bonding is strong enough to have significant effects on physical properties, resulting in some unexpected properties for water.

The solid (ice) is less dense than the liquid (water)

Solids are usually denser than liquids – but ice is less dense than water.

- Particles in solids are **usually** packed closer together than in liquids.
- Hydrogen bonds hold water molecules apart in an open lattice structure.

Therefore ice is less dense than water.

When water changes state, the covalent bonds between the H and O atoms in an H_2O molecule are strong and do **not** break.

It is the intermolecular forces that break.

The diagram below shows how the open lattice of ice collapses on melting.

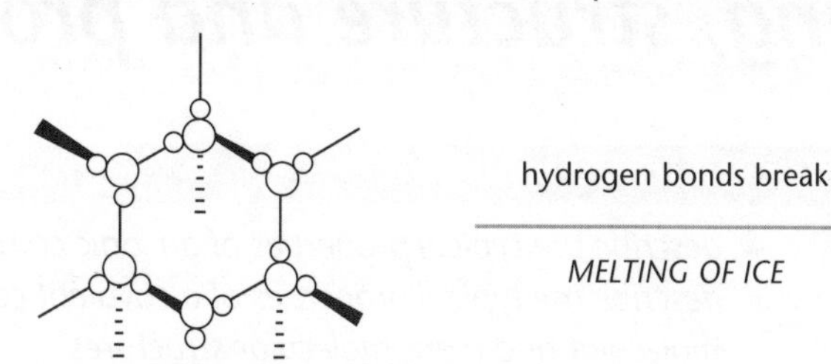

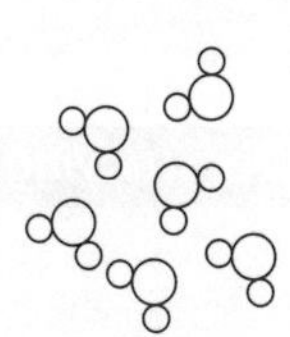

tetrahedral open lattice in ice

ice lattice collapses: molecules move closer together

Ice has a relatively high melting point, and water a relatively high boiling point

- There are relatively strong hydrogen bonds between H_2O molecules.
- The hydrogen bonds are extra forces over and above van der Waals' forces.
- These extra forces result in higher melting and boiling points than would be expected from just van der Waals' forces.
- When the ice lattice breaks, hydrogen bonds are broken.

Other properties

The extra intermolecular bonding from hydrogen bonds also explains the relatively high surface tension and viscosity in water.

Evidence for hydrogen bonds

The boiling points of the Group 6 and Group 7 hydrides are shown below.

Group 6		*Group 7*	
hydride	*boiling point / K*	*hydride*	*boiling point / °C*
H_2O	373	HF	293
H_2S	212	HCl	188
H_2Se	232	HBr	206
H_2Te	271	HI	238

The boiling points increase from $H_2S \longrightarrow H_2Te$ and from $HCl \longrightarrow HI$:

- the number of electrons in molecules increase
- van der Waals' forces increase
- more energy is required to break the intermolecular bonds to vaporise the hydrides.

Without hydrogen bonding, H_2O would be a gas at room temperature and pressure.

The boiling point of the first hydride in each group is higher than expected. This provides evidence that there are some extra forces acting between the molecules that must be broken to boil each hydride. These extra forces are hydrogen bonds.

Progress check

1 Which of the following molecules have hydrogen bonding: H_2O, H_2S, CH_4, CH_3OH, NO_2?

2 Draw diagrams showing hydrogen bonding between:
(a) 2 molecules of ammonia
(b) 1 molecule of water and 1 molecule of ethanol.

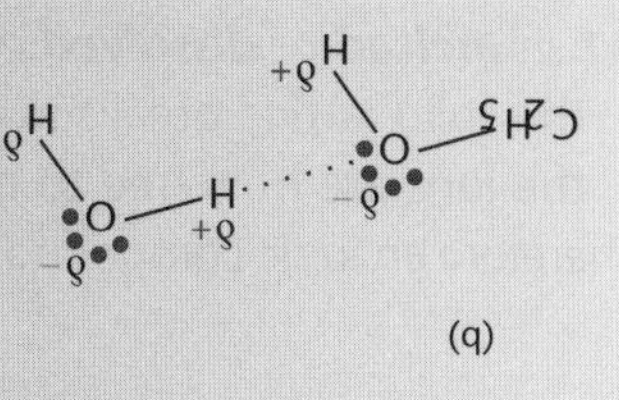

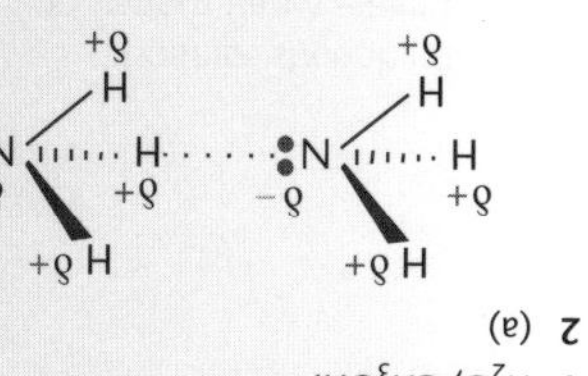

(b)

1 H_2O, CH_3OH.

2 (a)

2.5 Bonding, structure and properties

After studying this section you should be able to:

LEARNING SUMMARY

- *describe the typical properties of an ionic compound in terms of its structure*
- *describe the typical properties of a covalent compound in terms of simple molecular and giant molecular structures*
- *describe the structure and associated properties of diamond and graphite*
- *understand the properties of metals in terms of metallic bonding*

Bonds and forces

AQA M1

Bonding, structure and properties are all related.

A structure is held together by bonds and forces. The different types of bonds introduced in the last section are shown below.

These are very important and are needed to understand the links between bonding, structure and properties.

KEY POINT

Covalent bonds act between atoms.
Ionic bonds act between ions.
Metallic bonds act between positive ions and electrons.
Hydrogen bonds act between polar molecules.
Dipole-dipole interactions act between polar molecules.
Van der Waals' forces act between molecules.

Ionic, covalent and metallic bonds are of comparable strength.

Intermolecular forces are much weaker, and their relative strengths are compared with the strength of covalent bonds in the table below.

type of bond	*bond enthalpy / kJ mol^{-1}*
covalent bond	200–500
hydrogen bond	5–40
van der Waals' forces	~2

The properties of a substance depend upon its **bonding** and **structure**.

Properties of ionic compounds

AQA M1

Ionic compounds form giant ionic lattices with each ion surrounded by ions of the opposite charge. The ions are held together by **strong** electrostatic attraction between positive and negative ions. (See page 40.)

Ionic compounds are solids at room temperature.

The strong forces between ions result in high melting and boiling points.

High melting point and boiling point

- High temperatures are needed to break the strong electrostatic forces holding the ions rigidly in the solid lattice.

Therefore ionic compounds have high melting and boiling points.

Electrical conductivity

Ionic compounds conduct only when ions are free to move – when molten or in aqueous solution.

*In the **solid** lattice,*

- the ions are in a fixed position and there are no mobile charge carriers.

Therefore an ionic compound is a **non-conductor** of electricity in the solid state.

*When **melted** or **dissolved** in water,*

- the solid lattice breaks down
- the ions are now free to move as mobile charge carriers.

Therefore an ionic compound is a **conductor** of electricity in liquid and aqueous states.

Solubility

- The ionic lattice often dissolves in ***polar*** solvents (e.g. water).
- Polar water molecules break down the lattice and surround each ion in solution as shown below for sodium chloride.

The solubility of ionic compounds in water increases with increasing temperature.

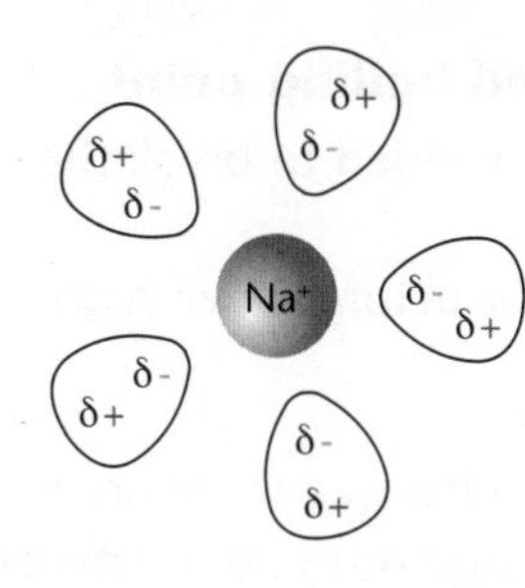

Na+ attracts δ– partial charges on the O atoms of water molecules.

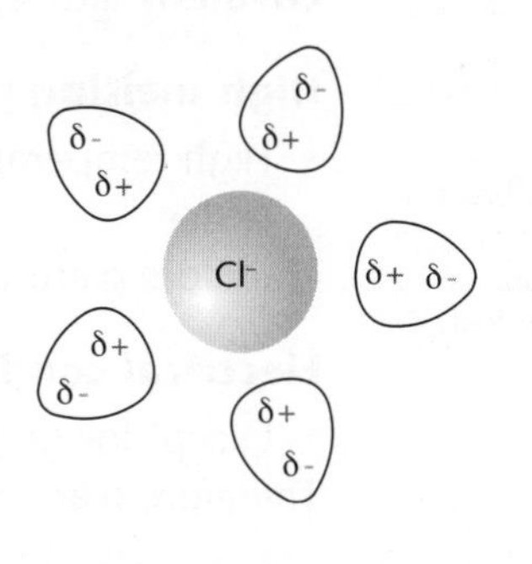

Cl– attracts δ+ partial charges on the H atoms of water molecules.

- Water molecules attract Na^+ and Cl^- ions.
- Lattice breaks down as it dissolves.
- Water molecules surround ions.

Properties of covalent compounds

AQA M1

Elements and compounds with covalent bonds have either of two structures:

- a simple molecular structure
- a giant covalent structure.

Intermolecular forces are weak.

Only the intermolecular forces break when a simple molecule melts or boils.

A common mistake in exams is to confuse intermolecular forces with covalent bonds.

Simple molecular structures, e.g. iodine, I_2

Simple molecular structures form solid lattices with small molecules, such as Ne, H_2, O_2, N_2, held together by **weak** intermolecular forces.

Low melting point and boiling point

- Low temperatures provide sufficient energy to break the weak intermolecular forces between molecules.

Therefore simple molecular structures have low melting and boiling points.

The simple molecular structure of solid I_2

Giant structure: high melting point.

Simple molecular structure: low melting point.

- When the I_2 lattice breaks down, only the weak van der Waals' forces between the I_2 molecules break.
- In the I_2 molecule, the covalent bond, I–I, is strong and does **not** break when the lattice breaks down.

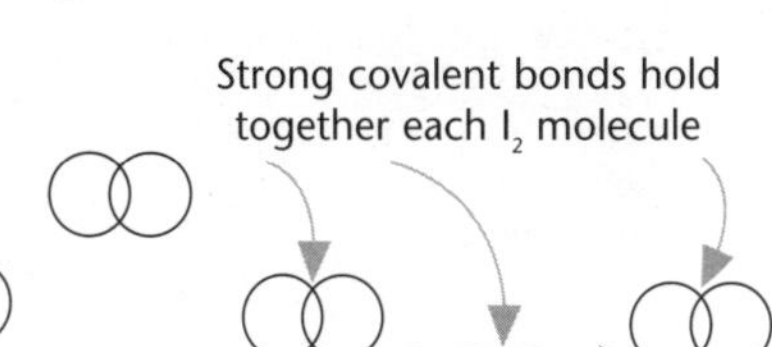

Electrical conductivity

- There are no mobile charged particles.

Therefore simple molecular structures are **non-conductors** of electricity.

Solubility

- Van der Waals' forces form between a simple molecular structure and a non-polar solvent, such as hexane. These weaken the structure.

Therefore simple molecular structures are often soluble in **non-polar** solvents (e.g. hexane).

Giant covalent structures, e.g. carbon (diamond and graphite)

Diamond, graphite and SiO_2 are examples of giant covalent lattices.

Giant covalent structures have thousands of atoms bonded together by **strong** covalent bonds. This type of structure is known by a variety of names: A giant **covalent** lattice, a giant **molecular** lattice or a giant **atomic** lattice.

Covalent bonds are strong.

The covalent bonds break when a giant molecular structure melts or boils – this only happens at high temperatures.

High melting point and boiling point

- High temperatures are needed to break the strong covalent bonds in the lattice.

Therefore giant covalent structures have high melting and boiling points.

Electrical conductivity

- Except for graphite (see below), there are no mobile charged particles.

Therefore giant covalent structures are **non-conductors** of electricity.

Carbon can also form other structures: fullerenes and nanotubes.

Solubility

- The strong covalent bonds in the lattice are too strong to be broken by either polar **or** non-polar solvents.

Therefore giant covalent structures are insoluble in polar ***and*** non-polar solvents.

Silicon (IV) oxide, SiO_2, has a similar tetrahedral structure to diamond.

Comparison between the properties of diamond and graphite

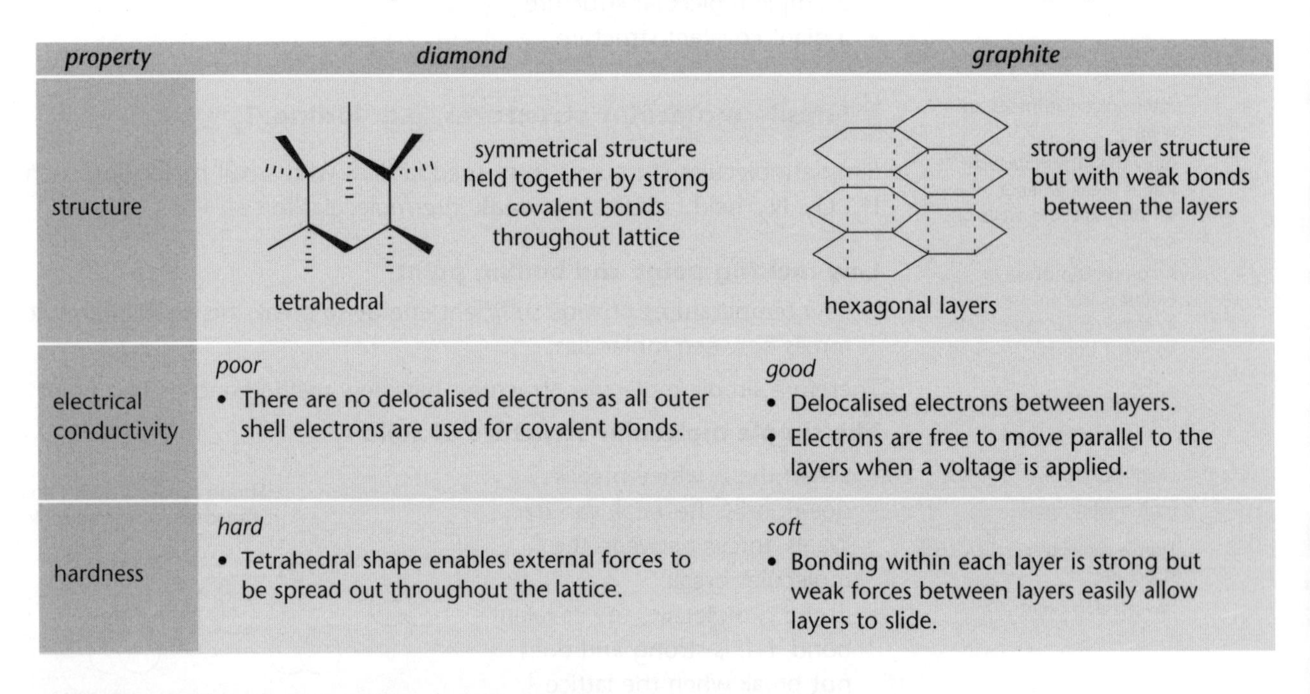

property	*diamond*	*graphite*
structure	tetrahedral symmetrical structure held together by strong covalent bonds throughout lattice	hexagonal layers strong layer structure but with weak bonds between the layers
electrical conductivity	*poor* • There are no delocalised electrons as all outer shell electrons are used for covalent bonds.	*good* • Delocalised electrons between layers. • Electrons are free to move parallel to the layers when a voltage is applied.
hardness	*hard* • Tetrahedral shape enables external forces to be spread out throughout the lattice.	*soft* • Bonding within each layer is strong but weak forces between layers easily allow layers to slide.

Properties of metals

AQA M1

Metals have a giant metallic lattice structures held together by **strong** electrostatic attractions between positive ions and negative electrons.

High melting point and boiling point

- Generally high temperatures are needed to separate the ions from their rigid positions within the lattice.

Therefore most metals have high melting and boiling points.

When a metal conducts electricity, only the electrons move.

Good thermal and electrical conductivity

- The existence of mobile, delocalised electrons allows metals to conduct heat and electricity well, even in the solid state.
- The electrons are free to flow between positive ions.
- The positive ions do **not** move.

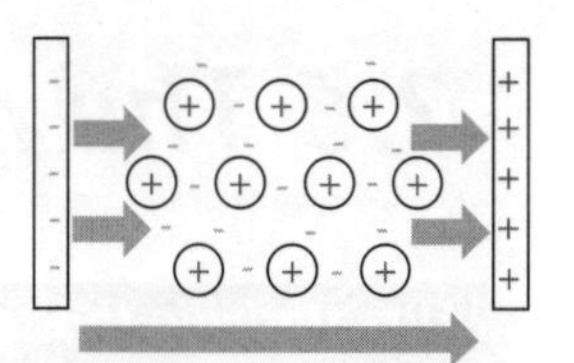

drift of delocalised electrons across potential difference

Summary of properties from structure and bonding

KEY POINT

A high melting point is the result of any giant structure, bonded together with strong forces. Giant structures can be ionic, covalent or metallic.

structure	*bonding*	*melting pt / boiling pt*	*reason*	*electrical conductivity*	*reason*	*solubility*	*reason*
giant ionic	ionic bonds throughout structure	high	strong electrostatic attraction between oppositely charged ions	poor when solid good when aqueous or molten	ions in a fixed position in lattice lattice has broken down: mobile ions	good in polar solvents, e.g. water	attraction between ionic lattice and polar solvent
simple molecular	covalent bonds **within** molecules, van der Waals' forces **between** molecules	low	weak van der Waals' forces between molecules	poor	no mobile charged particles (electrons or ions)	good in non-polar solvents	van der Waals' forces between molecular structure and solvent
giant covalent	covalent bonds throughout structure	high	strong covalent bonds between atoms	poor	no mobile charged particles (electrons or ions)	poor	forces within lattice too strong to be broken by solvents
hydrogen bonded	hydrogen bonds between molecules	low *but* higher than expected	dipole-dipole attractions between molecules	poor	no mobile charged particles (electrons or ions)	good in polar solvents, e.g. water	attraction between dipoles
giant metallic	metallic bonds throughout structure	usually high	strong electrostatic attractions between ions and electrons	good	mobile electrons, even in solid state	poor	forces within lattice too strong to be broken by solvents

Progress check

1 For (a) MgO; (b) CH_4; (c) SiO_2, state the structure and explain the following physical properties: melting and boiling points, electrical conductivity, solubility.

1 (a) giant ionic: high m.pt/b.pt. – strong forces between ions.
Non-conductor when solid – ions fixed in lattice. Conductor when molten or dissolved in water – ions are able to move.
Soluble in water – dipole on water attracted to ions.
(b) simple molecular: low m. pt/b. pt. – weak van der Waals' forces between molecules.
Non-conductor – no mobile charge carriers, electrons localised in covalent bonds.
Soluble in non-polar solvents – van der Waals' forces in solvent interact with van der Waals' forces in CH_4.
(c) giant molecular: high m. pt/b. pt. – strong covalent bonds between atoms.
Non-conductor – no mobile charge carriers, electrons localised in covalent bonds.
Insoluble in all solvents – strong covalent bonds are too strong to be broken by either polar or non-polar solvents.

2.6 The modern Periodic Table

After studying this section you should be able to:

- *describe the Periodic Table in terms of atomic number, periods and groups*
- *classify the Periodic Table into s, p, d and f blocks*
- *describe and explain periodic trends in atomic radii, ionisation energies, melting and boiling points, and electrical conductivities*

LEARNING SUMMARY

Arranging the elements

AQA M1

In the Periodic Table:

- Elements are arranged in order of increasing atomic number
- Groups are vertical columns including elements with similar properties
- Periods are horizontal rows across which there is a trend in properties.

Key areas in the Periodic Table

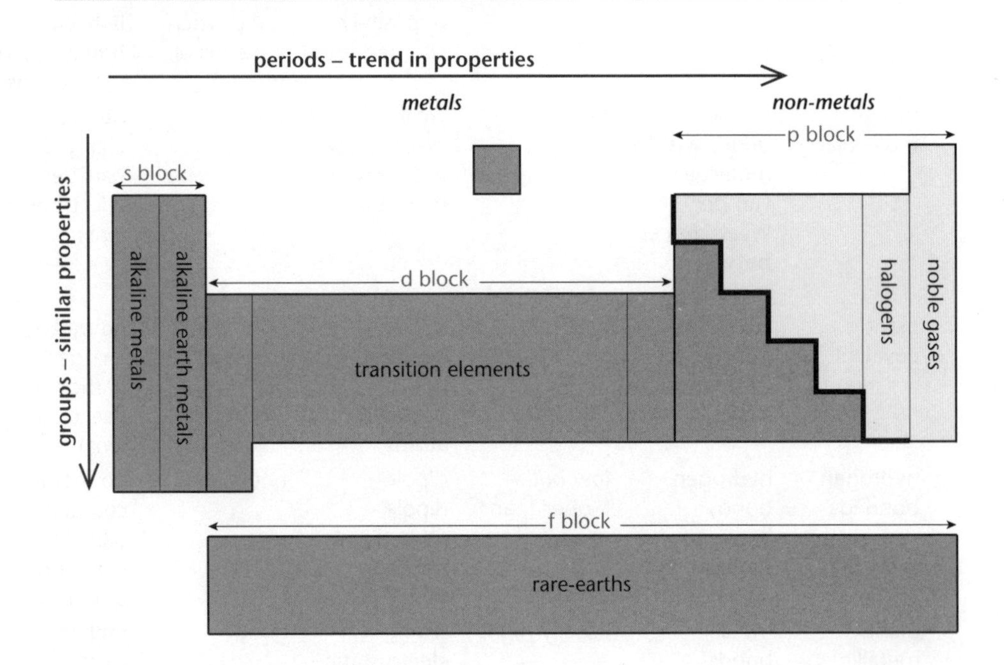

- The diagonal stepped line separates metals (to the left) from non-metals (to the right).
- Elements close to this line, such as silicon and germanium, are called **semi-metals** or **metalloids**. They show properties intermediate between those of a metal and a non-metal.
- The four blocks (s, p , d and f) show the sub-shell being filled.

Periodicity

The trend in properties across a period is repeated across each period – this is called *periodicity*.

Periodicity is the periodic trend in properties, repeated across each period.

e.g.	Period 2	METAL	⟶	NON-METAL
	Period 3	METAL	⟶	NON-METAL

This periodicity of properties means that predictions can be made about the likely properties of an element and its compounds.

Group 4
C
Si
Ge
Sn
Pb
non-metal
To metal

Note, however, the metal/non-metal divide. On descending the Periodic Table, the changeover from metal to non-metal takes place further to the right. For example at the top of Group 4 carbon is a non-metal whereas, at the bottom of Group 4, tin and lead are metals. This means that subtle trends in properties take place down a group.

Trends in atomic radii and first ionisation energies

AQA M1

Ionisation energies measure the ease with which an atom of an element loses electrons (see page 19).

Across a period

Across a period, the atomic radius decreases. The trend in atomic radius across Period 2 is shown below:

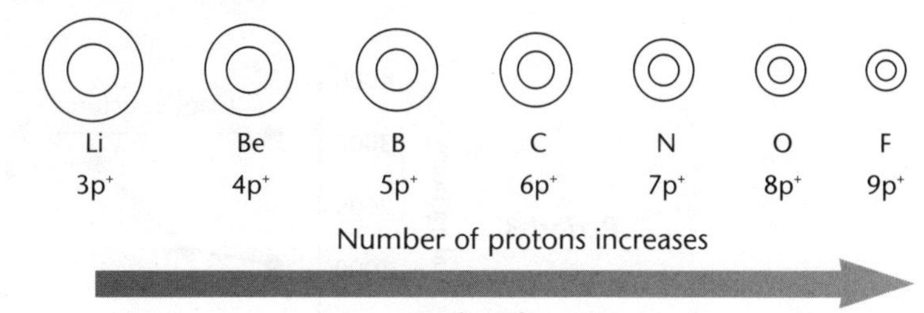

Atomic radius decreases across a period.

First ionisation energy increases across a period.

Across a period:

- the **nuclear charge increases**
- outer electrons are being added to the same shell
- the attraction between the nucleus and outer electrons increases
- the atomic radius **decreases**
- the first ionisation energy **increases**.

Down a group

Atomic radius increases down a group.

First ionisation energy decreases down a group.

Down a group, the nuclear charge also increases but this is more than compensated by the increase in atomic radius and shielding.

Down a group:

- extra shells are added that are **further** from the nucleus
- there are more shells between the outer electrons and the nucleus leading to **greater shielding** of the nuclear charge
- the attraction between the nucleus and the outer electrons decreases
- the atomic radius **increases**
- the first ionisation energy **decreases**.

Shells increase
Shielding increases
Atomic radius increases

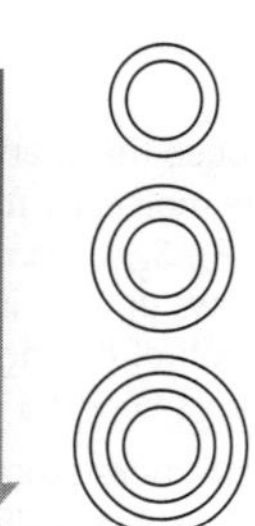

KEY POINT

Across a Period, nuclear charge is the key factor.

Down a Group, atomic radius and shielding are the key factors (see also ionisation energies, page 19).

Trends in melting and boiling points

AQA M1

Trends in melting and boiling point provide information about structure: Giant or simple molecular.

The graphs below show the variation in boiling points across Period 2 and Period 3 of the Periodic Table.

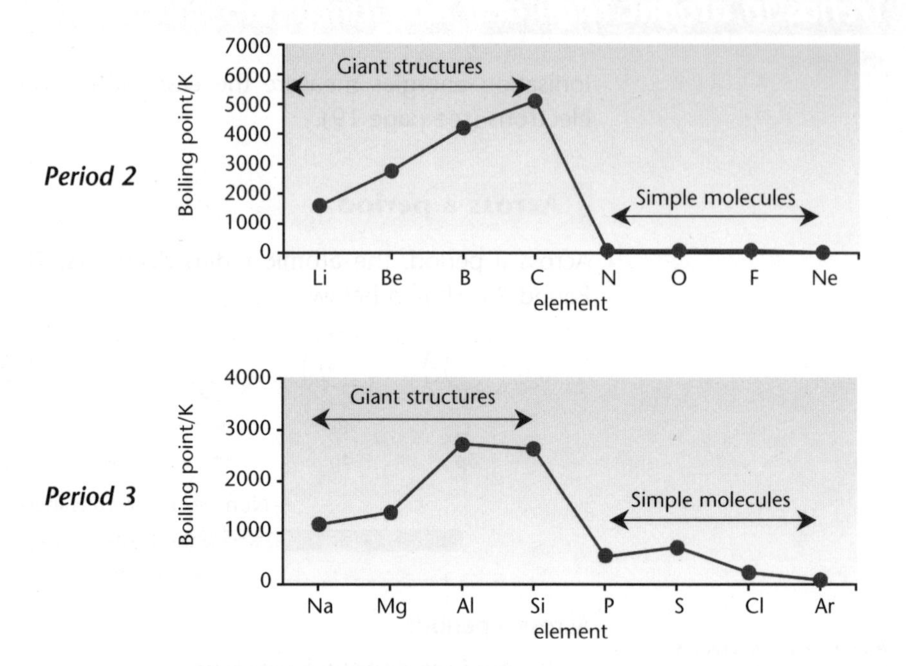

Across a period:

- the boiling point increases from Group 1 to Group 4
- there is a sharp decrease in boiling point between Group 4 and Group 5
- the boiling points are comparatively low from Group 5 to Group 8.

KEY POINT

The sharp decrease in boiling point marks a change from giant structures to simple molecular structures.

Note the periodicity in boiling point: The trend for Period 2 is repeated across Period 3. The trend in melting points is similar.

Notice the changes in the molecular formulae of P_4, S_8, Cl_2 and Ar. This shows up well in the graph of boiling points across Period 3.

With more atoms (and consequently electrons) in each molecule, the forces between molecules (van Der Waals' forces) increase.

More details of the structures of these elements are shown below.

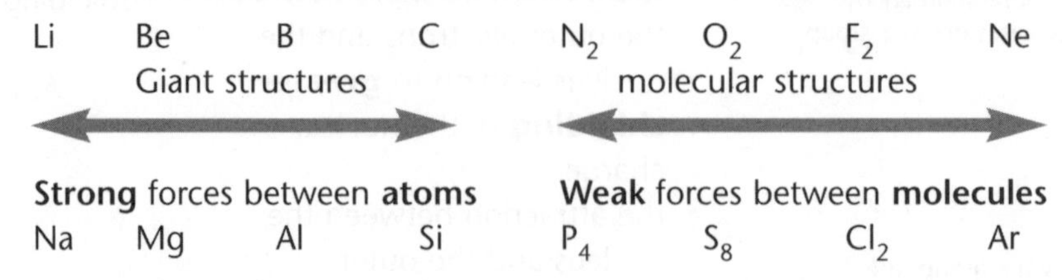

When a giant structure is melted or boiled:

- strong forces are broken
- a large input of energy is required
- the melting and boiling points are high.

When a simple molecular structure is melted or boiled:

- weak forces between molecules are broken
- a relatively small input of energy is required
- the melting and boiling points are low.

Comparison of the boiling points of the metals

Note the increase in boiling points of the metals Na → Al in Period 3.

- The number of delocalised electrons in the lattice increases.
- The charge on each cation increases.
- This results in increasing attractive forces within the metallic lattice.

The diagram below compares the attractive forces between atoms of sodium, magnesium and aluminium.

See also Metallic bonding, pages 44, 54–55.

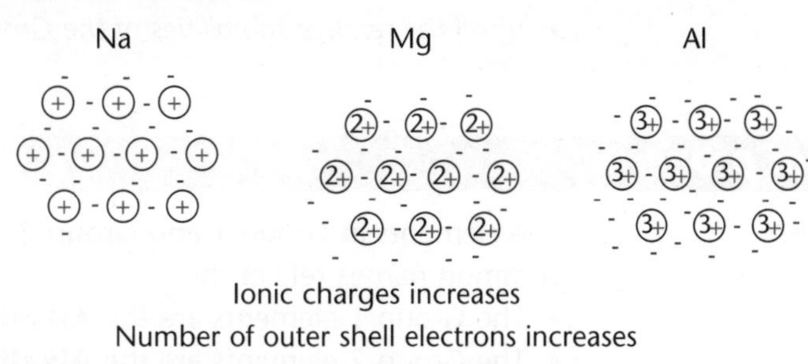

Attraction increases: Boiling point increases

The increasing number of delocalised electrons in the lattice also explains the increase in electrical conductivity from Na → Al.

Progress check

1 Using ideas about nuclear charge, attraction and shells, explain the trend in atomic radii across a period and down each group.

2 Why is the melting point of carbon much higher than that of nitrogen?

1 Period: The nuclear charge increases as electrons are being added to the same shell. The attraction between the nucleus and outer electrons increases ∴ The atomic radius decreases across a period.
Group: Extra shells are added that are further from the nucleus leading to an increased shielding of the outer electrons from the nucleus. Both these factors lead to less attraction between the nucleus and the outer electrons ∴ The atomic radius increases down a group.

2 Carbon has a giant molecular structure; between the atoms there are strong covalent bonds which break on melting. Nitrogen has a simple molecular structure; between the molecules there are weak van Der Waals' forces that break on melting.

2.7 The s-block elements: Group 1 and Group 2

After studying this section you should be able to:

- *describe the characteristic properties of the s-block elements*
- *describe the reactions of the s-block elements with oxygen and with water*
- *recall some of the reactions of s-block oxides*
- *recall the relative solubilities of the Group 2 hydroxides and sulfates.*

LEARNING SUMMARY

General properties

AQA M2

The elements in Group 1 and Group 2 have hydroxides that are alkaline and their common names reflect this.

- The Group 1 elements are the **Alkali Metals**.
- The Group 2 elements are the **Alkaline Earth Metals**.

Electron configuration

The elements in Group 1 and Group 2 have their highest energy electrons in an s sub-shell and these two groups are known as the *s-block elements*.

Each Group 1 element has:

- **one** electron more than the electron configuration of a noble gas
- an outer s sub-shell containing **one** electron.

Each Group 1 element reacts in a similar way as each atom has 1 electron in the outer shell.

The s-block elements
Group 1
Group 2

Electron configuration of the s-block elements

Group 1		Group 2	
Li	[He] 2s1	Be	[He] 2s2
Na	[Ne] $3s^1$	Mg	[Ne] $3s^2$
K	[Ar] $4s^1$	Ca	[Ar] $4s^2$
Rb	[Kr] $5s^1$	Sr	[Kr] $5s^2$
Cs	[Xe] $6s^1$	BA	[Xe] $6s^2$
Fr	[Rn] $7s^1$	Ra	[Rn] $7s^2$

Each Group 2 element has:

- **two** electrons more than the electron configuration of a noble gas
- an outer s sub-shell containing **two** electrons.

Each Group 2 element reacts in a similar way as each atom has 2 electrons in the outer shell.

Physical properties

Group 1

- They are soft metals and can be cut with a knife.
- They have low melting and boiling points.
- They have low densities: Li, Na and K all float on water.
- They have colourless compounds.

Densities of s-block elements

Group 1 Density/g cm^{-3}		Group 2 Density/g cm^{-3}	
Li	0.53	Be	1.85
Na	0.97	Mg	1.74
K	0.86	Ca	1.54
		Sr	2.60
		BA	3.51

Densities for comparison:
H_2O, 1.00 g cm^{-3}; Fe: 7.86 g cm^{-3}

Group 2

- They have reasonably high melting and boiling points.
- They have low densities although not as low as those in Group 1.
- They have colourless compounds.

Melting points of s-block elements			
Group 1 Melting point/° C		Group 2 Melting point/° C	
Li	181		
Na	98	Mg	649
K	63	Ca	839
		Sr	769
		BA	725

Melting points for comparison: Fe: 1535°C; Cu 1083°C

Melting points show a general decrease down Group 1 and Group 2:

- the atoms increase in size
- the outer electrons are further away from the nucleus
- the attractive force of nuclei on the outer electrons decreases.

Reactivity of the s-block elements

 M2

The elements in the s-block are the most reactive metals and are strong reducing agents.

Li	Be	
Na	Mg	
K	Ca	Reactivity
Rb	Sr	**Increases**
Cs	BA	
Fr	Ra	

Down Group 1 and Group 2

Atomic radius increases

First ionisation energy decreases

See page 57 for a detailed explanation.

Group 1 elements are oxidised in reactions, each atom losing one electron from its outer s sub-shell to form a 1+ ion (+1 oxidation state):

$$M \longrightarrow M^{+} + e^{-}$$

Group 2 elements are oxidised in reactions, each atom losing two electrons from its outer s sub-shell to form a 2+ ion (+2 oxidation state):

$$M \longrightarrow M^{2+} + 2e^{-}$$

Reactivity **increases** down each group reflecting the increasing ease of losing electrons. The ionisation energy of the metal is an important factor in this process.

> **KEY POINT**
>
> Within each group, the elements become **more** reactive as the group descends.
>
> First ionisation energy **decreases** down the group.

The Group 2 elements

AQA M2

Reaction with oxygen

The Group 2 elements react vigorously with oxygen. Each element forms the expected ionic oxide with the general formula, $M^{2+}O^{2-}$:

e.g. $$2Ca(s) + O_2(g) \longrightarrow 2CaO(s)$$

Action of water

Reactions of Group 2 elements with oxygen and water are redox reactions.

Mg forms MgO with steam.

Reactivity increases down the group reflecting the increasing ease with which electrons can be lost.

- Mg reacts very slowly with water, forming the **hydroxide** and hydrogen:

$$Mg(s) + 2H_2O(l) \longrightarrow Mg(OH)_2(aq) + H_2(g)$$

- With steam, reaction is much quicker forming the **oxide** and hydrogen:

$$Mg(s) + H_2O(g) \longrightarrow MgO(s) + H_2(g)$$

- Further down the group from calcium, each metal reacts vigorously with water:

$$Ca(s) + 2H_2O(l) \longrightarrow Ca(OH)_2(aq) + H_2(g)$$

Group 2 oxides and hydroxides

$Mg(OH)_2$, $Ca(OH)_2$, $Sr(OH)_2$, $Ba(OH)_2$ — solubility increases, alkalinity increases (down the group)

Thus solid barium hydroxide is reasonably soluble in water to form a strong alkaline solution with greater $OH^-(aq)$ concentration.

Reaction with water

The Group 2 oxides form alkaline solutions with water:

e.g. $$MgO(s) + H_2O(l) \longrightarrow Mg^{2+}(aq) + 2OH^-(aq)$$

- Magnesium hydroxide, $Mg(OH)_2(s)$ is only slightly soluble in water and so the resulting solution is only a dilute alkali.
- The solubility in water increases down the group and resulting solutions are more alkaline:

$$Ba(OH)_2(s) + aq \longrightarrow Ba^{2+}(aq) + 2OH^-(aq)$$

The alkalinity of the Group 2 hydroxides is exploited commercially. Some indigestion tablets contain $Mg(OH)_2$ to neutralise excess acid in the stomach. Farmers add $Ca(OH)_2$ as 'lime' to neutralise acid soils.

Reaction with acids

- The Group 2 oxides behave as bases and are neutralised by acids, such as $HCl(aq)$, forming salts and water:

e.g. $$MgO(s) + 2HCl(aq) \longrightarrow MgCl_2(aq) + H_2O(l)$$

Solubility of Group 2 sulfates

In medicine, barium sulfate is used as a 'barium meal' which shows up imperfections in the gut when exposed to X-rays. Barium compounds are extremely poisonous in solution but the insolubility of $BaSO_4$ is such that no harm is caused to the patient by this treatment.

Magnesium sulfate $MgSO_4$ is very soluble in water but the solubility decreases as the group is descended. The trend is so marked that barium sulfate $BaSO_4$ is virtually insoluble.

In the laboratory, the precipitation of $BaSO_4$ is used to test for the sulfate ion. A solution of a soluble barium salt (usually $BaCl_2(aq)$ or $Ba(NO_3)_2$) is added to a solution of a substance in dilute nitric acid. In the presence of aqueous sulfate ions, a dense white precipitate of barium sulfate is formed.

$$Ba^{2+}(aq) + SO_4^{2-}(aq) \longrightarrow BaSO_4(s)$$

$MgSO_4$, $CaSO_4$, $SrSO_4$, $BaSO_4$ — solubility **decreases** (down the group)

Progress check

1 Write down equations for the following reactions:
(a) barium with water (b) calcium oxide with nitric acid.

2 Identify the oxidation number changes taking place during the thermal decomposition of sodium nitrate:

$$2NaNO_3(s) \longrightarrow 2NaNO_2(s) + O_2(g)$$

1 (a) $Ba(s) + 2H_2O(l) \longrightarrow Ba(OH)_2(aq) + H_2(g)$
(b) $CaO(s) + 2HNO_3(aq) \longrightarrow Ca(NO_3)_2(aq) + H_2O(l)$

2 N, $+5 \longrightarrow +3$; O, $-2 \longrightarrow 0$.

2.8 The Group 7 elements and their compounds

After studying this section you should be able to:

- *describe the characteristic properties of the Group 7 elements*
- *recall the relative reactivity of the halogens as oxidising agents*
- *recall the characteristic tests for halide ions*
- *describe the reactions of halides with concentrated sulfuric acid*
- *describe the use of thiosulfate titrations.*

LEARNING SUMMARY

General properties

AQA M2

The common name for the elements in Group 7 is the **halogens**.

Electron configuration

Each halogen has **seven** outer shell electrons; just one electron short of the electron configuration of a noble gas. The outer **p** sub-shell contains **five** electrons.

Electron configuration of the halogens
F [He] $2s^22p^5$
Cl [Ne] $3s^23p^5$
Br [Ar] $3d^{10}4s^24p^5$
I [Kr] $4d^{10}5s^25p^5$
At [Xe] $4f^{14}5d^{10}6p^26p^5$

Trend in physical states

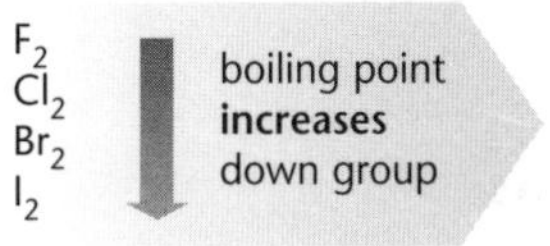

The halogens exist as diatomic molecules, X_2. The boiling points of the halogens increase on descending the group.

- The physical states of the halogens at r.t.p. show the classic trend of gas ⟶ liquid ⟶ solid.
- On descending the group the number of electrons increases leading to an increase in van der Waals' forces between molecules.
- Therefore the boiling point increases.

F_2, Cl_2, Br_2, I_2: boiling point **increases** down group

Boiling points of the Halogens

	Boiling point/°C	State at r.t.p.
F_2	–188	gas
Cl_2	–35	gas
Br_2	59	liquid
I_2	184	solid
At_2	337	solid

Trend in electronegativity

AQA M2

Electronegativity is a measure of the attraction of an atom for the pair of electrons in a covalent bond (see also page 47).

The hydrogen halides have polar molecules: $H^{\delta+}–X^{\delta-}$. The polarity decreases on descending the halogens and the order of polarity is:

most polar H–F > H–Cl > H–Br > H–I *least polar*

This trend in polarity results from the **decreasing** electronegativity of the halogen atom on descending the group.

- The atomic radius increases from F → Cl → I resulting in less nuclear attraction on the bonding electrons (this is despite the increase in nuclear charge!).
- There are more electron shells between the nucleus and the bonding electrons to shield the nuclear charge.

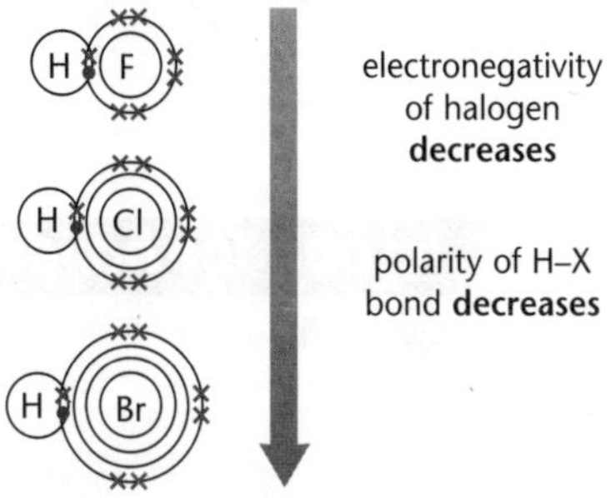

H F, H Cl, H Br: electronegativity of halogen **decreases**; polarity of H–X bond **decreases**

Electronegativity of the Halogens	
	Pauling value
F	4.0
Cl	3.0
Br	2.8
I	2.5

The overall effect is that the smaller the halogen atom, the greater the nuclear attraction experienced by the bonding electrons. Thus the large electronegativity of fluorine results in a bond between hydrogen and fluorine that is more polar than between hydrogen and other halogens.

The relative reactivity of the halogens as oxidising agents

AQA M2

> The halogens are the most reactive non-metals and are strong oxidising agents.
>
> The halogens become **less** reactive as the group descends as their oxidising power decreases.
>
> KEY POINT

The halogens are reduced in reactions, each atom gaining one electron into a p sub-shell to form a –1 ion (–1 oxidation state):

e.g. $\frac{1}{2}F_2(g)$ + e^- ⟶ $F^-(g)$

[He] $2s^22p^5$ ⟶ [He] $2s^22p^6$ (or [Ne])

- An extra electron is captured by being attracted to the outer shell of an atom by the nuclear charge of an atom.

F, Cl, Br, I, At: reactivity as oxidising agent **decreases**

On descending the halogens:

- the atomic radii increase resulting in less nuclear attraction at the edge of the atom (despite the increase in nuclear charge).
- there are more electron shells between the nucleus and the edge of the atom to shield the nuclear charge.

F

Cl

Br

The overall effect is that most nuclear attraction is experienced at the edge of the small fluorine atoms. This attraction decreases down the group as the atoms get bigger.

Therefore the oxidising power of the halogens decreases down the group. Fluorine is the strongest oxidising agent and is able to attract an extra electron more strongly than other halogens.

Displacement reactions of the halogens

The decrease in reactivity as the group is descended can be demonstrated by displacement reactions of aqueous halides using Cl_2, Br_2 and I_2.

chlorine oxidises both Br^- and I^-:

$$Cl_2(aq) + 2Br^-(aq) \longrightarrow 2Cl^-(aq) + Br_2(aq)$$

$$Cl_2(aq) + 2I^-(aq) \longrightarrow 2Cl^-(aq) + I_2(aq)$$

bromine oxidises I^- only:

$$Br_2(aq) + 2I^-(aq) \longrightarrow 2Br^-(aq) + I_2(aq)$$

iodine does not oxidise either Cl^- or Br^-

The formation of halogens in displacement reactions is identified by colours, which are more distinctive in organic solvents:

halogen	*water*	*hexane*
Cl_2	pale green	pale green
Br_2	orange	orange
I_2	brown	purple

Industrial extraction of bromine

The main source of bromine is as bromide ions, Br^- in sea water. Bromine is extracted by oxidising sea water with chlorine. Because chlorine is a stronger oxidising agent, it displaces the bromide ions using the principles of the displacement reaction.

Testing for halide ions

AQA M2

Addition of aqueous silver ions (using $AgNO_3(aq)$) to a solution of halide ions in dilute nitric acid produces coloured precipitates that have different solubilities in aqueous ammonia.

fluoride: $Ag^+(aq) + F^-(aq) \not\longrightarrow$ no precipitate

chloride: $Ag^+(aq) + Cl^-(aq) \longrightarrow AgCl(s)$ white precipitate, soluble in dilute $NH_3(aq)$

bromide: $Ag^+(aq) + Br^-(aq) \longrightarrow AgBr(s)$ cream precipitate, soluble in conc. $NH_3(aq)$

iodide: $Ag^+(aq) + I^-(aq) \longrightarrow AgI(s)$ yellow precipitate, insoluble in conc. $NH_3(aq)$

In sunlight, silver halides are reduced to silver. This reaction formed the basis of old photographic films.

Reactions of halides with concentrated sulfuric acid

AQA M2

Concentrated sulfuric acid is an oxidising agent. Halide salts react with concentrated sulfuric acid producing a range of products depending on the halide used. This difference is caused by the increasing reducing power of the hydrogen halides as the group is descended.

HCl, HBr, HI — increased strength as reducing agents

NaCl and H_2SO_4

Hydrogen chloride gas, HCl, is formed.

$$NaCl(s) + H_2SO_4(l) \longrightarrow NaHSO_4(s) + HCl(g)$$

HCl does **not** reduce H_2SO_4

The HCl formed is not a sufficiently strong reducing agent to reduce the sulfuric acid. No redox reaction takes place.

NaBr and H_2SO_4

Hydrogen bromide gas, HBr, is initially formed.

$$NaBr(s) + H_2SO_4(l) \longrightarrow NaHSO_4(s) + HBr(g)$$

Some of the hydrogen bromide reduces the sulfuric acid with the formation of sulfur dioxide and orange bromine fumes.

$$2HBr(g) + H_2SO_4(l) \longrightarrow SO_2(g) + Br_2(g) + 2H_2O(l)$$

$+6 \xrightarrow{-2} +4$ *S reduced*

$2 \times -1 \xrightarrow{+2} 0$ *Br oxidised*

HBr reduces H_2SO_4

$H_2SO_4 \xrightarrow{HBr} Br_2 + SO_2$

NaI and H_2SO_4

Hydrogen iodide gas, HI, is initially formed.

$$NaI(s) + H_2SO_4(l) \longrightarrow NaHSO_4(s) + HI(g)$$

Hydrogen iodide is a strong reducing agent and reduces the sulfuric acid to a mixture of reduced products including sulfur dioxide and hydrogen sulfide.

Reduction to SO_2 (ox no: +4)

$$2HI(g) + H_2SO_4(l) \longrightarrow SO_2(g) + I_2(s) + 2H_2O(l)$$

$+6 \xrightarrow{-2} +4$ *S reduced*

$2 \times -1 \xrightarrow{+2} 0$ *I oxidised*

Further reduction to H_2S (ox no: −2)

$$6HI(g) + SO_2(g) \longrightarrow H_2S(g) + 3I_2(s) + 2H_2O(l)$$

$+4 \xrightarrow{-6} -2$ *S reduced*

$6 \times -1 \xrightarrow{+6} 0$ *I oxidised*

HI reduces H_2SO_4 and SO_2

$H_2SO_4 \xrightarrow{HI} I_2 + SO_2$

$SO_2 \xrightarrow{HI} I_2 + H_2S$

Disproportionation

AQA M2

Disproportionation is a reaction in which the same element is both oxidised and reduced.

Disproportionation of chlorine in water

Chlorine reacts with water. This reaction is an example of *disproportionation* in which chlorine is both reduced (to chloride, Cl^-) and oxidised (to chlorate(I) ClO^-).

$$Cl_2(aq) + H_2O(l) \longrightarrow HClO(aq) + HCl(aq)$$

$0 \xrightarrow{-1} -1$ *chlorine reduced*

$0 \xrightarrow{+1} +1$ *chlorine oxidised*

A solution of chlorine in water is pale green, showing the presence of chlorine.

A small amount of chlorine is added to drinking water to kill bacteria that would make the water unsafe to drink. It has been claimed that chlorine treatment of water has done more to improve public health than any other treatment, preventing diseases such as cholera and typhoid. This is despite the toxicity of chlorine and possible risks from the formation of toxic chlorohydrocarbons by reaction with organic matter.

Disproportionation of chlorine in dilute aqueous alkalis

Dilute aqueous sodium hydroxide reacts with halogens when mixed at **room temperature**.

This reaction is another example of *disproportionation* in which chlorine is both reduced (to chloride, Cl^-) and oxidised (to chlorate(I) ClO^-).

$$Cl_2(aq) + 2NaOH(aq) \longrightarrow NaCl(aq) + NaClO(aq) + H_2O(l)$$

$$0 \xrightarrow{\;-1\;} -1 \qquad \textit{chlorine reduced}$$

$$0 \xrightarrow{\;+1\;} +1 \qquad \textit{chlorine oxidised}$$

The solution formed is the basis for common bleach.

In some areas fluoride ions are added to drinking water to reduce tooth decay.

Some people believe that fluoride ions pose other detrimental health risks and that the rights of the individual to use or reject fluoride have been compromised.

Progress check

1 State and explain the trend in boiling points of the halogens fluorine to iodine.

2 How could you distinguish between NaCl, NaBr and NaI by a simple test?

3 Comment on the changes in oxidation number of chlorine in the following reaction: $Cl_2(aq) + H_2O(l) \longrightarrow HClO(aq) + H^+(aq) + Cl^-(aq)$

1 Boiling points increase down the group. From F_2 to I_2 the number of electrons increase leading to greater van der Waals' forces between molecules and higher boiling points.

2 Add $AgNO_3$(aq). NaCl gives a white precipitate, soluble in dilute ammonia. NaBr gives a cream precipitate, soluble in concentrated ammonia. NaI gives a yellow precipitate, insoluble in concentrated ammonia.

3 $Cl_2 \longrightarrow HClO$, Cl: $0 \longrightarrow +1$ (oxidation). $Cl_2 \longrightarrow Cl^-$, Cl: $0 \longrightarrow -1$ (reduction). Chlorine has been both oxidised and reduced (disproportionation).

2.9 Extraction of metals

After studying this section you should be able to:

LEARNING SUMMARY

- *understand that carbon reduction is used to extract iron from its ore*
- *understand that electrolysis is used to extract aluminium from its ore*
- *understand that metal reduction is used to extract titanium from its ore*
- *understand the importance of economic factors and recycling*

Reduction of metals

AQA M2

The main methods for extracting metals from their ores are:

- reduction of the ore with carbon
- reduction of the molten ore by electrolysis
- reduction of the ore with a more reactive metal.

Reduction of metal oxides with carbon

AQA M2

Production of iron

The main ore of iron, haematite, contains Fe_2O_3 and this is reduced by coke in the blast furnace. This is a continuous process in which high quality haematite, coke and limestone are fed in at the top of the furnace and hot air blown in near the bottom.

Coke contains a very high carbon content. Coke is cheap and provides an economic source of carbon and carbon dioxide as reducing agents.

The production of iron in the blast furnace is a **continuous** process. Raw materials are added at the top of the furnace and the products removed at the base. For further production, more raw materials are added and the process continues in a continuous flow, sometimes for months on end.

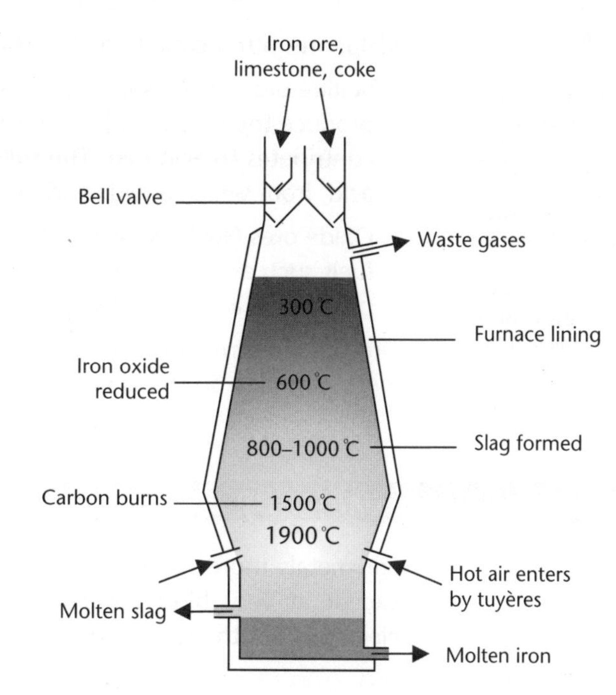

- Initially, hot coke reacts with air forming carbon monoxide:

$$2C(s) + O_2(g) \longrightarrow 2CO(g)$$

- The carbon monoxide reduces most of the iron oxide at around 1200°C:

$$Fe_2O_3(s) + 3CO(g) \longrightarrow 2Fe(l) + 3CO_2(g)$$

- In hotter parts of the furnace, coke also reacts directly with iron oxide:

$$2Fe_2O_3(s) + 3C(s) \longrightarrow 4Fe(l) + 3CO_2(g)$$

This equation represents the overall reaction for the reduction.

Both C and CO are reductants in the blast furnace.

Limestone removes SiO_2.

The limestone is used to remove acidic impurities such as silicon dioxide (sand). This produces a slag, largely of calcium silicate, $CaSiO_3$:

$$CaCO_3 + SiO_2 \longrightarrow CaSiO_3 + CO_2$$

- Molten iron collects at the bottom of the furnace and is run off as 'pig iron.'
- Pig iron is very impure and brittle, containing about 4% of carbon (as well as Mn, Si, P and S).

Production of other metals

Reduction with carbon is a common method for obtaining many other metals from their ores.

- Manganese

  Pyrolusite ore contains manganese(IV) oxide. The ore is heated with carbon to obtain manganese by reduction:

$$MnO_2 + 2C \longrightarrow Mn + 2CO$$
$$MnO_2 + 2CO \longrightarrow Mn + 2CO_2$$

- Copper

  Copper is usually reduced from its sulfide ores. It can also be obtained by high temperature reduction using carbon:

$$CuO + C \longrightarrow Cu + CO$$
$$CuO + CO \longrightarrow Cu + CO_2$$

Environmental pollution problems:

sulfide ores form sulfur dioxide;

reduction with carbon produces greenhouse gases.

Aluminium oxide is very stable and could only be reduced by carbon at very high temperatures. Electrolysis is used instead to extract aluminium (see page 69).

Limitations of carbon reduction

Metal ores often contain metal oxides or sulfides.

- Sulfide ores (such as galena, PbS, and sphalerite, ZnS) are first roasted in air to produce the metal oxide. The sulfur is oxidised as sulfur dioxide, which contributes to acid rain. The sulfur dioxide can be reacted to make sulfuric acid, from which a wide variety of useful products can be manufactured.
- Oxide ores (such as haematite, Fe_2O_3) are reduced to the metal using fossil fuels such as coke from coal. Carbon dioxide, emitted as the main gaseous product, is a greenhouse gas.

Ores of some metals (such as titanium and tungsten) react with carbon forming metal carbides, so this is not a practical method for extracting these metals.

$$TiO_2 + 3C \longrightarrow TiC + 2CO$$

Reduction of metal oxides by electrolysis of the molten ore

AQA M2

For metals more reactive than zinc, reduction with carbon does not take place except at very high temperatures and these metals are usually extracted by electrolysis of the molten ore.

The extraction of aluminium

The melting point of aluminium oxide is 2045°C but this is decreased using molten cryolite. This reduces the energy requirements.

The main ore of aluminium, bauxite, contains Al_2O_3. Purified bauxite is dissolved in molten cryolite (Na_3AlF_6) at 970°C. This is a continuous process needing regular additions of aluminium oxide.

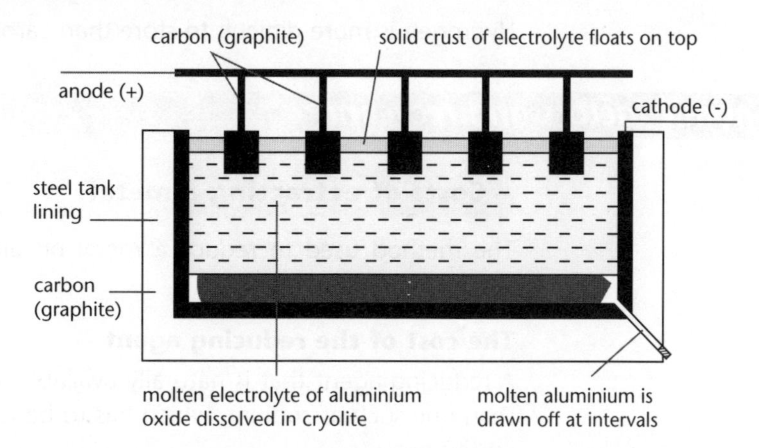

The electrodes are made of graphite and the cell reactions are:

at the cathode $Al^{3+} + 3e^- \longrightarrow Al$

at the anode $2O^{2-} \longrightarrow O_2 + 4e^-$

- Periodically the graphite anodes need replacing because, at the high temperatures used, the carbon anodes react with the liberated oxygen.
  $2C + O_2 \longrightarrow 2CO$
  and $C + O_2 \longrightarrow CO_2$
- The process consumes large amounts of electricity (as electricity is needed to melt the Al_2O_3 as well as to reduce it).
- The process is only economic where electricity is relatively inexpensive.

Reduction of metal halides with metals

AQA M2

Some metals, such as titanium, become brittle if contaminated with traces of impurities such as carbon, oxygen or nitrogen.

Highly pure metals can be formed by reduction of a metal halide with a reactive metal.

Despite the high cost, pure titanium is manufactured by reduction of its molten chloride with a more reactive metal.

When pure, titanium is a light metal with a high strength and high resistance to corrosion.

When impure, it is brittle and of little use.

Production of titanium

The main ore of titanium, rutile, contains titanium(IV) oxide. Titanium is manufactured in a two-stage batch process.

- Rutile is first converted to titanium(IV) chloride using chlorine and coke at around 900 °C:
  $$TiO_2 + 2C + 2Cl_2 \longrightarrow TiCl_4 + 2CO$$
- The titanium(IV) chloride is purified from other chlorides (e.g. those of iron, silicon and chromium) by fractional distillation under argon or nitrogen.
- The chloride is then reduced by a more reactive metal such as sodium or magnesium:
  $$TiCl_4 + 4Na \longrightarrow Ti + 4NaCl$$
  $$TiCl_4 + 2Mg \longrightarrow Ti + 2MgCl_2$$
- An inert atmosphere of argon is used to prevent any contamination of the metal with oxygen or nitrogen.
- If sodium has been used, the sodium chloride by-product is washed out with dilute hydrochloric acid, leaving titanium as a granular powder.
- If magnesium has been used, the magnesium chloride by-product is removed by distillation at high temperature and low pressure.

The production of titanium is a **batch** process. Raw materials are reacted together in a single 'batch' and the products are separated from the reaction mixture. For further production, the whole process needs to be repeated from scratch.

Reduction of metal oxides with hydrogen

AQA M2

The main ore of tungsten wolframite, is processed to precipitate tungsten(VI) oxide, WO_3. The tungsten(VI) oxide is reduced with hydrogen gas:

$$WO_3 + 3H_2 \longrightarrow W + 3H_2O$$

Hydrogen is more difficult to store than carbon and is highly flammable.

Economic factors and recycling

AQA M2

Costs of extracting a metal

The method used to reduce a metal on an industrial scale is dependent upon several factors:

The cost of the reducing agent

A reducing agent that is naturally available such as carbon (from coal) is cheaper than one such as sodium, which has to be prepared first by a separate (and often costly) process.

The energy costs for the process

The lower the temperature of a process, the lower the energy requirement and the cheaper the process. Carbon reduction is a high temperature process with substantial energy costs.

Instead of using high temperatures to extract copper, scrap iron can be added to solutions containing copper ions. The more reactive iron displaces copper from solution.

This process also allows copper to be extracted from low grade ore, containing small proportions of copper.

The purity required of the metal

It is relatively expensive to produce a metal with very high purity. There needs to be a high demand for the metal if this extra expense is to be justified.

These factors need to be weighed against each other when considering the overall cost of an extraction process.

Recycling

Metals are a valuable resource and, instead of disposal, metals are often collected and recycled.

Recycling of iron

Recycling metals conserves metal ores and saves energy.

- Iron is recycled by collection of scrap iron which is melted and reused.
- The Earth's reserves of iron ore are conserved.
- The energy required to mine iron ore, transport it and smelt it is several times greater than the energy required to recycle scrap iron.

Recycling of aluminium

The cost of re-using scrap aluminium is only one-twentieth of the cost of making the pure metal.

- Owing to its very high resistance to corrosion, used aluminium is as good as new.
- Recycling conserves the Earth's supply of aluminium.
- The electrolysis of aluminium oxide requires high temperatures and vast quantities of electricity.
- It is also cheaper and easier to refine recycled aluminium than to mine bauxite and extract the metal by electrolysis.

Sample question and model answer

Chemical bonding helps to explain different properties of materials.

The phrase 'electrostatic attraction between ions' is essential when describing an ionic bond.

(a) Using suitable diagrams, explain what is meant by *ionic*, *covalent* and *metallic* bonding.

An ionic bond is the electrostatic attraction between ions. ✓
An example is sodium chloride:

DO use diagrams in your answers. Here, dot and cross diagrams are essential for showing ionic and covalent compounds.

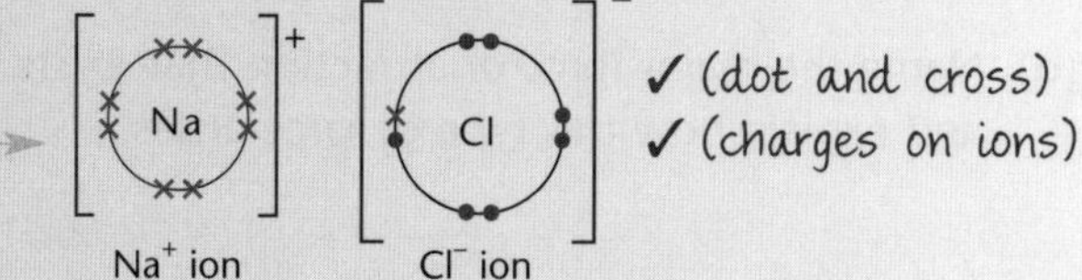

The word 'pair' is essential when describing a covalent bond. Omit it and you risk losing the mark.

A covalent bond is a **shared** ✓ **pair** of electrons. ✓
An example is hydrogen chloride.

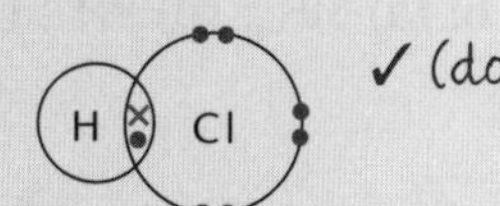

Metallic bonding occurs between positive centres/ions ✓ surrounded by mobile or delocalised electrons: ✓

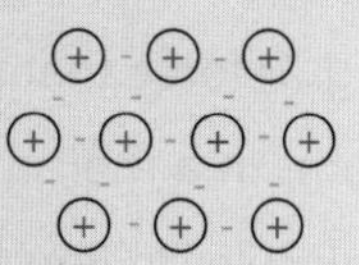

Metallic bonding is the attraction between positive ions and mobile electrons. ✓

There are plenty of examples. Keep them simple and make sure that you choose correctly. Don't use NaCl as a covalent example – it is ionic!

Many candidates do actually make this mistake.

(b) Three substances have ionic, covalent and metallic bonding respectively. Compare and explain the electrical conductivity of the three materials.

The ionic compound does not conduct electricity when solid ✓ because the ions are fixed ✓ in a lattice.
The ionic compound does conduct electricity when aqueous or molten ✓ because the ions are mobile. ✓
The covalent compound does not conduct ✓ at all because there are no free charge carriers ✓ (electrons or ions).
The metallic compound does conduct electricity ✓ because the delocalised electrons are able to move ✓ across a potential difference.

17 marking points ⟶ maximum of [15]

Special care needed throughout with language. A-grade students will score all of these marks.

Ionic bonding: positive ions and negative ions – **both** carry electricity.

Metallic bonding: positive ions and delocalised electrons – **only** the electrons move and carry electricity.

Practice examination questions

1 (a) A water molecule, H_2O, is bonded by covalent bonds. What is meant by the term *covalent bond*? [2]

(b) Use the formation of the H_3O^+ ion from water to explain what is meant by a *dative covalent bond*. [2]

(c) State the bond angle in a water molecule and predict, with an explanation, the bond angle in an H_3O^+ ion. [4]

(d) Name the major force of attraction that exists between molecules in water and explain how this type of force arises. [3]

[Total: 11]

2 Describe how *electron pair repulsion* explains the shapes of simple molecules. Show, using diagrams, how electron pair repulsion determines the molecular shapes and bond angles in molecules of (a) SiH_4; (b) PH_3; (c) BCl_3. [13]

[Total: 13]

3 When liquid bromine, Br_2, is heated gently, it forms an orange–brown vapour. When a crystal of potassium chloride is heated to the same temperature, it does not change state.

(a) Name the type of bonding or force that occurs:

(i) between bromine atoms in a molecule of bromine

(ii) between bromine molecules in liquid bromine. [2]

(b) Explain why liquid bromine turns into a vapour when heated gently. [1]

(c) Explain, in terms of its bonding, why the crystal of potassium chloride does not melt or vaporise when heated gently. [2]

(d) Describe what happens to the particles in potassium chloride when the solid is heated above room temperature but below its melting point. [1]

(e) Suggest why much more energy is required to vaporise potassium chloride than to melt it. [2]

[Total: 8]

4 The boiling points of (a) water, (b) hydrogen chloride and (c) krypton are shown below.

liquid	H_2O	HCl	Kr
boiling point /°C	100	–85	–152

Suggest reasons for the different boiling points by considering the nature and strength of the intermolecular forces in each case.

[Total: 12]

Practice examination questions

5 In the Periodic Table, describe and explain the trend in the atomic radii of the elements in Period 3 and in Group 2. [Total: 8]

6 Barium and magnesium are elements in Group 2 of the Periodic Table.

(a) Complete and balance the following equations:

(i) $Ba(s) + H_2O(l) \longrightarrow$

(ii) $Mg(s) + O_2(g) \longrightarrow$

(iii) $Mg(s) + HCl(aq) \longrightarrow$ [3]

(b) (i) Suggest why the reactivity of the Group 2 elements increases on descending the group.

(ii) Name one reaction of Group 2 elements that illustrates this trend of increasing reactivity. [4]

(c) (i) What is the property of magnesium oxide that makes it suitable for its use as a lining in some furnaces?

(ii) Name **two** other Group 2 compounds, and state a use for each of them. [3]

[Total: 10]

7 This question is about the Group 7 elements: chlorine, bromine and iodine.

(a) Describe and explain the trend in oxidising ability shown by the Group 7 elements. [5]

(b) Chlorine reacts with hot concentrated sodium hydroxide as in the equation below.

$$3Cl_2(g) + 6NaOH(aq) \longrightarrow 5NaCl(aq) + NaClO_3(aq) + 3H_2O(l)$$

(i) Use changes in oxidation numbers to show that this is a redox reaction.

(ii) Calculate the maximum mass of $NaClO_3$ that could be prepared from the reaction of 65 dm^3 of chlorine with hot concentrated sodium hydroxide at room temperature and pressure.

(iii) Chlorine reacts with **dilute** aqueous sodium hydroxide at room temperature. Write an equation for this reaction and state what the resulting solution could be used for. [9]

[Total: 14]

8 (a) (i) What does the term *electronegativity* mean?

(ii) What is the trend in electronegativity of the halogens fluorine to iodine? Explain why this trend happens. [5]

(b) State and explain the trend in volatility of the halogens fluorine to iodine. [3]

(c) Aqueous bromine was added separately to aqueous solutions of potassium chloride and potassium iodide. Describe what would be observed and write equation(s) for any reaction(s) that take place. [4]

(d) State and explain the trend in reactivity shown by the experiments in part (c). [2]

[Total: 14]

Chapter 3

Energetics, rates and equilibrium

The following topics are covered in this chapter:

- *Enthalpy changes*
- *Determination of enthalpy changes*
- *Bond enthalpy*
- *Reaction rates*
- *Catalysis*
- *Chemical equilibrium*

3.1 Enthalpy changes

LEARNING SUMMARY

After studying this section you should be able to:

- *understand that reactions can be exothermic or endothermic*
- *construct a simple enthalpy profile diagram for a reaction*
- *explain and use the terms: standard conditions, enthalpy changes of reaction, formation and combustion*

Energy out, energy in

AQA M2

Enthalpy, *H*, is the **heat energy** that is stored in a chemical system.

Enthalpy cannot be measured experimentally. However, an **enthalpy change** can be measured from the temperature change in a chemical reaction.

> **KEY POINT**
> **An enthalpy change** ΔH is the heat energy exchange with the surroundings at constant pressure.

The conservation of energy is an important principle in science. This is often summarised by the first law of thermodynamics.

The first law of thermodynamics underpins all of this section.

> **KEY POINT**
> **The first law of thermodynamics** states that energy may be exchanged between a chemical system and the surroundings but the *total* energy remains constant.

Exothermic reactions

During an exothermic reaction, heat energy is **released** to the surroundings.

Any energy **loss** from the chemicals is balanced by the same energy **gain** to the surroundings, which rise in temperature.

exothermic
chemicals lose energy: ΔH –ve
surroundings gain energy and rise in temperature: ΔT +ve

Energy pathway diagram (reaction profile)

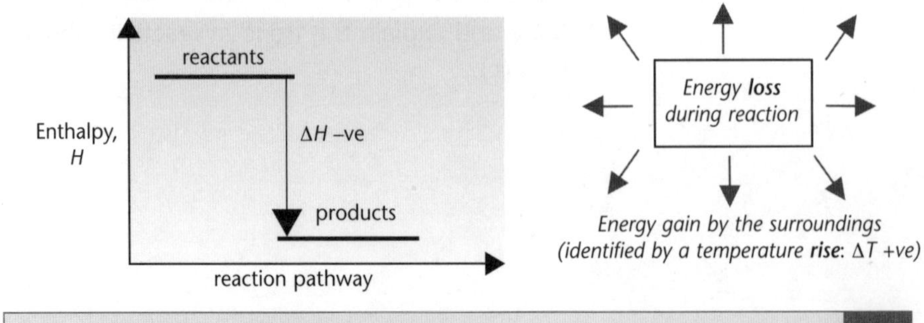

Chemists refer to *surroundings* as anything other than the reacting chemicals.

Surroundings often just means the water in which chemicals are dissolved.

> **KEY POINT**
> In an exothermic reaction, ΔH is negative:
> - heat is given out (**to** the surroundings)
> - the reacting chemicals lose energy.

Endothermic reactions

During an endothermic reaction, heat energy is **taken in** from the surroundings.

Any energy **gain** to the chemicals is provided by the same energy **loss** from the surroundings, which fall in temperature.

endothermic
chemicals gain energy: ΔH +ve
surroundings lose energy and fall in temperature: ΔT –ve

Energy pathway diagram (reaction profile)

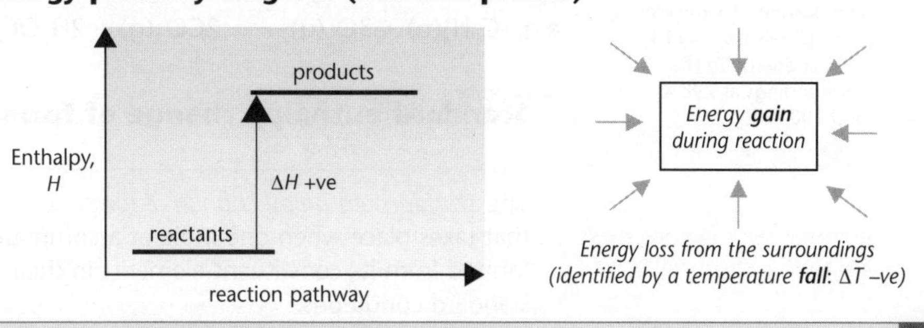

> In an endothermic reaction, ΔH is positive:
> - heat is taken in (**from** the surroundings)
> - chemicals gain energy.
>
> KEY POINT

Standard enthalpy changes

AQA M2

Enthalpy changes have been measured for many reactions. Many are recorded in data books as *standard enthalpy changes* and these are discussed below.

Standard conditions

$\Delta H^{\ominus}$ refers to an enthalpy (H) change (Δ) under standard conditions ($^{\ominus}$).

Standard pressure is 100 kPa or 1 bar.
The former standard pressure of 101 kPa or 1 atmosphere is still quoted in many books.

> *Standard conditions* are:
> - a pressure of 100 kPa
> - a stated temperature: 298 K (25°C) is usually used
> - a concentration of 1 mol dm^{-3} (*for aqueous solutions*).
>
> A *standard state* is the physical state of a substance under standard conditions.
>
> KEY POINT

The standard state of water at 298 K and 100 kPa is a liquid.

Standard enthalpy change of reaction

Always use a chemical equation or an unambiguous definition with a stated enthalpy change.

> The *standard enthalpy change of reaction* $\Delta H^{\ominus}_r$ is the enthalpy change that accompanies a reaction in the molar quantities that are expressed in a chemical equation under standard conditions, all reactants and products being in their standard states.
>
> KEY POINT

The enthalpy change of reaction $\Delta H^{\ominus}_r$ depends upon the quantities shown in a chemical equation. $\Delta H^{\ominus}_r$ should always be quoted with an equation.

$\Delta H^{\ominus}_r$ only has a meaning with an equation.
$\Delta H^{\ominus}_r$ has units of kJ mol^{-1}.
mol^{-1} means 'for the amount (in moles) shown in the equation'.

For the reaction:

$$H_2(g) \quad + \quad \tfrac{1}{2}O_2(g) \longrightarrow H_2O(l) \qquad \Delta H^{\ominus}_r = -286 \text{ kJ mol}^{-1}$$
1 mol   ½ mol   1 mol

but with twice the quantities, there is twice the enthalpy change:

$$2H_2(g) \quad + \quad O_2(g) \longrightarrow 2H_2O(l) \qquad \Delta H^{\ominus}_r = -572 \text{ kJ mol}^{-1}$$
2 mol   1 mol   2 mol

Standard enthalpy change of combustion

> **KEY POINT**
>
> The *standard enthalpy change of combustion* $\Delta H^{\ominus}_c$ is the enthalpy change that takes place when one mole of a substance reacts completely with oxygen under standard conditions, all reactants and products being in their standard states.

This means that complete combustion of 1 mole of $C_2H_4(g)$ releases 1411 kJ of heat energy to the surroundings at 298 K and 100 kPa.

e.g. $C_2H_4(g) + 3O_2(g) \longrightarrow 2CO_2(g) + 2H_2O(l)$ $\quad \Delta H^{\ominus}_c = -1411 \text{ kJ mol}^{-1}$

Standard enthalpy change of formation

> **KEY POINT**
>
> The *standard enthalpy change of formation* $\Delta H^{\ominus}_f$ is the enthalpy change that takes place when one mole of a compound in its standard state is formed from its constituent elements in their standard states under standard conditions.

This means that the formation of 1 mole of $H_2O(l)$ from 1 mole of $H_2(g)$ and ½ mole of $O_2(g)$ releases 286 kJ of heat energy to the surroundings at 298 K and 100 kPa.

e.g. $H_2(g) + \frac{1}{2}O_2(g) \longrightarrow H_2O(l)$ $\quad \Delta H^{\ominus}_f = -286 \text{ kJ mol}^{-1}$

For an element, the standard enthalpy change of formation is defined as zero.

The formation of $H_2(g)$ from $H_2(g)$ does not involve a chemical change so there is no enthalpy change.

Progress check

1 Draw energy profile diagrams for the following reactions:

(a) $N_2O_4(g) \longrightarrow 2NO_2(g)$ $\quad \Delta H = +58 \text{ kJ mol}^{-1}$

(b) $N_2(g) + 3H_2(g) \longrightarrow 2NH_3(g)$ $\quad \Delta H = -92 \text{ kJ mol}^{-1}$

2 Write an equation to represent the enthalpy change for:

(a) $\Delta H^{\ominus}_c(CH_4)$; (b) $\Delta H^{\ominus}_f(NO_2)$

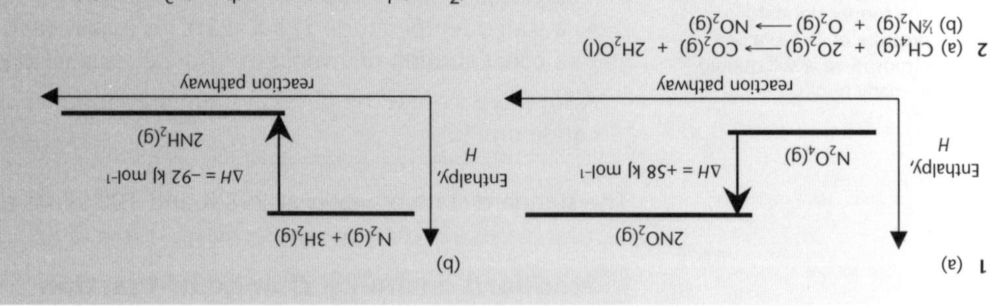

3.2 Determination of enthalpy changes

After studying this section you should be able to:

LEARNING SUMMARY

- *calculate enthalpy changes from direct experimental results, using the relationship: energy change = $mc\Delta T$*
- *use Hess' Law to construct enthalpy cycles*
- *determine enthalpy changes indirectly, using enthalpy cycles and enthalpy changes of formation and combustion*

Direct determination of enthalpy changes

AQA M2

Calculating enthalpy changes

The **heat energy change**, ***Q***, in the surroundings can be calculated using the relationship below.

$$Q = mc\Delta T \text{ Joules}$$

- *m* is the **mass** of the surroundings that experience the temperature change
- *c* is the **specific heat capacity** of the surroundings
- ΔT is the **temperature change** (final temperature – initial temperature)

The **specific heat capacity** of a substance is the energy required to raise the temperature of 1 g of a substance by 1°C.

Example

Addition of an excess of magnesium to 100 cm^3 of 2.00 mol dm^{-3} $CuSO_4(aq)$ raised the temperature from 20.0°C to 65.0°C. Find the enthalpy change for the reaction:

$$Mg(s) + CuSO_4(aq) \longrightarrow MgSO_4(aq) + Cu(s)$$

specific heat capacity of solution $c = 4.18$ J g^{-1} K^{-1}
density of solution = 1.00 g cm^{-3}.

Assume that solids, such as Mg(s) make little difference to the energy change.

Find the heat energy change

100 cm^3 of solution has a mass of 100 g;
Temperature change, ΔT = (65.0–20.0)°C = +45.0°C
Heat energy **gain** to surroundings, $Q = mc\Delta T$ = 100 × 4.18 × 45.0 J = **+18810 J**
∴ heat energy **loss** from the reacting chemicals = **–18810 J**

Any energy **gain** by the surroundings must have come from the same energy **loss** in the chemical reaction.

Find out the amount (in moles) that reacted

Amount (in mol) of $CuSO_4$ that reacted = $2.00 \times \frac{100}{1000}$ mol = 0.200 mol

Scale the quantities to those in the equation

Mg(s) +	$CuSO_4(aq)$	⟶	$MgSO_4(aq)$	+	Cu(s)
1 mol	1 mol		1 mol		1 mol

For 0.200 mol (1/5th mol) of $CuSO_4$, $\Delta H = -18810$ J
For 1 mol of $CuSO_4$, $\Delta H = 5 \times -18810 = -94050$ J
$\Delta H = -94.1$ kJ mol^{-1} (to 3 sig. figs)

∴ enthalpy change of reaction is given by:

$$Mg(s) + CuSO_4(aq) \longrightarrow MgSO_4(aq) + Cu(s) \quad \Delta H = -94.1 \text{ kJ mol}^{-1}$$

All values in this example are to 3 significant figures. This indicates the expected accuracy of the answer which should also be expressed to 3 significant figures.

In this case, $\Delta H = -94.1$ kJ mol^{-1}.

Indirect determination of enthalpy changes

AQA M2

Hess' Law

Many reactions have enthalpy changes that cannot be found directly from a single experiment. Hess' Law provides a method for the *indirect* determination of enthalpy changes.

Hess' Law is an extension of the First Law of Thermodynamics.

> *Hess' Law* states that, if a reaction can take place by more than one route and the initial and final conditions are the same, the total enthalpy change is the same for each route.
>
> KEY POINT

The diagram below shows two routes for converting reactants into products.

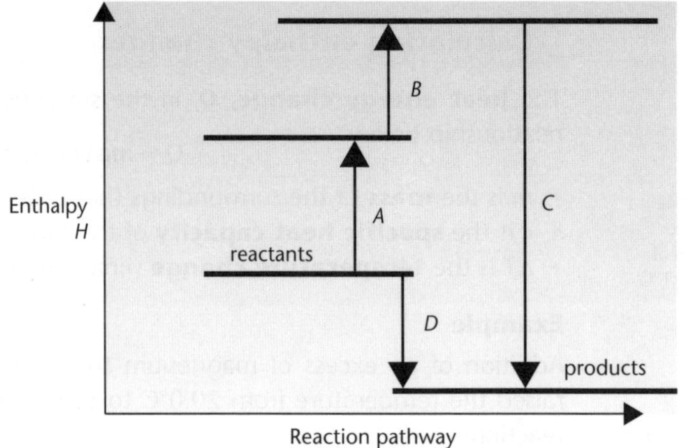

Indirect determination of an enthalpy change uses an energy cycle based on Hess' Law.

This method is used when the reaction is very difficult to carry out and related reactions can be measured more easily.

Following the arrows from reactants to products,

Route 1: **A + B + C**

Route 2: **D**

By Hess' Law, the total enthalpy change is the same for each route.

∴ **A + B + C = D**

If three of these enthalpy changes are known, the fourth can always be calculated.

Enthalpy changes of combustion are required for all reactants and products.

Using $\Delta H^{\ominus}_c$ values to determine an enthalpy change indirectly

Example:

Find the enthalpy change for the reaction:

$$C(s) + 2H_2(g) \longrightarrow CH_4(g)$$

substance	*C(s)*	*$H_2(g)$*	*$CH_4(g)$*
$\Delta H^{\ominus}_c$ / kJ mol^{-1}	–394	–286	–890

Always show your working. In exams, you are rewarded for a good method.

If you make one small slip in a calculation, you may only lose 1 mark provided that you show clear working.

Use the $\Delta H^{\ominus}_c$ data as a 'link' to construct an energy cycle.

- An energy cycle is constructed by linking the reactants and products to their **combustion products**.
- Note the direction of the arrows **from** the reactants and products of the reaction **to** the common combustion products.

The common combustion products here are $CO_2(g)$ and $H_2O(l)$. You can include these in your cycle but they are not required and 'combustion products' have been used here.

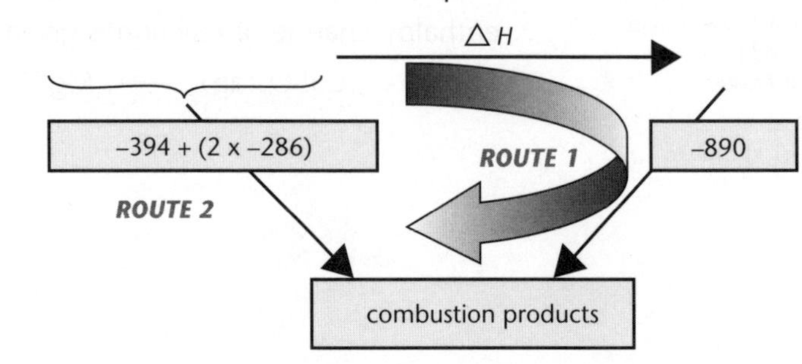

Calculate the unknown enthalpy change

By Hess' Law,

$$\text{Route 1: } \Delta H + [(-890)]$$
$$\text{Route 2: } [(-394) + (2 \times -286)]$$
$$\therefore \underbrace{\Delta H + [(-890)]}_{\text{Route 1}} = \underbrace{[(-394) + (2 \times -286)]}_{\text{Route 2}}$$
$$\Delta H = [(-394) + (2 \times -286)] - [(-890)]$$
$$\Delta H = \mathbf{-76\ kJ\ mol^{-1}}$$
$$\therefore C(s) + 2H_2(g) \longrightarrow CH_4(g) \qquad \Delta H^{\ominus} = -76\ kJ\ mol^{-1}$$

For enthalpy changes of combustion data only,
$\Delta H = \Sigma\Delta H^{\ominus}_c$ (reactants) – $\Sigma\Delta H^{\ominus}_c$ (products)

KEY POINT

Enthalpy changes of formation are required for all reactants and products that are compounds.
For elements, $\Delta H^{\ominus}_f = 0$ (formation of the element from the element – no change).

Using $\Delta H^{\ominus}_f$ values to determine an enthalpy change indirectly

Example:

Find the enthalpy change for the reaction:

$$C_2H_6(g) + 3\tfrac{1}{2}O_2(g) \longrightarrow 2CO_2(g) + 3H_2O(l)$$

substance	$C_2H_6(g)$	$CO_2(g)$	$H_2O(l)$
$\Delta H^{\ominus}_f$/ kJ mol^{-1}	–85	–394	–286

Use the $\Delta H^{\ominus}_f$ data as a 'link' to construct an energy cycle

- An energy cycle is constructed linking the reactants and products with their **constituent elements**.
- Note the direction of the arrows **from** the constituent elements to the reactants and products of the reaction.

Notice that $\Delta H^{\ominus}(O_2) = 0$ kJ mol^{-1}
This can be omitted from the cycle.

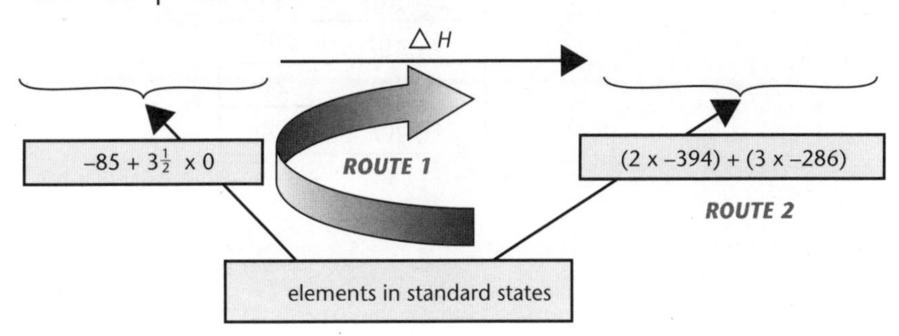

Calculate the unknown enthalpy change

By Hess' Law,

$$\text{Route 1: } [(-85) + (3\tfrac{1}{2} \times 0)] + \Delta H$$
$$\text{Route 2: } [(2 \times -394) + (3 \times -286)]$$
$$\therefore \underbrace{[(-85) + 0] + \Delta H}_{\textit{Route 1}} = \underbrace{[(2 \times -394) + (3 \times -286)]}_{\textit{Route 2}}$$
$$\therefore \Delta H = [(2 \times -394) + (3 \times -286)] - [(-85) + 0]$$
$$\therefore \Delta H = \mathbf{-1561}\ kJ\ mol^{-1}$$
$$C_2H_6(g) + 3\tfrac{1}{2}O_2(g) \longrightarrow 2CO_2(g) + 3H_2O(l) \qquad \Delta H^{\ominus} = -1561\ kJ\ mol^{-1}$$

For enthalpy changes of formation data only,
$\Delta H = \Sigma\Delta H^{\ominus}_f$ (products) – $\Sigma\Delta H^{\ominus}_f$ (reactants)

KEY POINT

Progress check

For questions 1 and 2 assume that:

specific heat capacity of solution $c = 4.18\ \text{J g}^{-1}\ \text{K}^{-1}$;

density of solution $= 1.00\ \text{g cm}^{-3}$.

1 Addition of zinc powder to 55.0 cm^3 of aqueous copper(II) sulfate at 22.8 °C raised the temperature to 32.3 °C. 0.3175 g of copper were obtained.
(a) Calculate the energy released in this reaction.
(b) Write an equation, including state symbols, for this reaction.
(c) Calculate the enthalpy change for this reaction per mole of copper formed.

2 Combustion of 1.60 g of ethanol, C_2H_5OH, raised the temperature of 150 g of water from 22.0 °C to 71.0 °C. Find the enthalpy change of combustion of ethanol.

3 Use the $\Delta H^{\ominus}_{c}$ data below to calculate enthalpy changes for:
(a) $3C(s) + 4H_2(g) \longrightarrow C_3H_8(g)$
(b) $C(s) + 2H_2(g) + \frac{1}{2}O_2(g) \longrightarrow CH_3OH(l)$

substance	$\Delta H^{\ominus}_{c}$ / kJ mol^{-1}
$C(s)$	−394
$H_2(g)$	−286
$C_3H_8(g)$	−2219
$CH_3OH(l)$	−726

4 Use the $\Delta H^{\ominus}_{f}$ data below to calculate enthalpy changes for:
(a) $C_2H_4(g) + H_2(g) \longrightarrow C_2H_6(g)$
(b) $SO_2(g) + 2H_2S(g) \longrightarrow 2H_2O(l) + 3S(s)$

compound	$\Delta H^{\ominus}_{f}$ / kJ mol^{-1}
$C_2H_4(g)$	+52
$C_2H_6(g)$	−85
$SO_2(g)$	−297
$H_2S(g)$	−21
$H_2O(l)$	−286

1 (a) 2.18 kJ; (b) $Zn(s) + CuSO_4(aq) \longrightarrow Cu(s) + ZnSO_4(aq)$; (c) −437 kJ mol^{-1}.
2 −883 kJ mol^{-1}.
3 (a) −107 kJ mol^{-1}; (b) −240 kJ mol^{-1}.
4 (a) −137 kJ mol^{-1}; (b) −233 kJ mol^{-1}.

3.3 Bond enthalpy

After studying this section you should be able to:

- *understand and use the term bond enthalpy*
- *explain chemical reactions in terms of enthalpy changes associated with the breaking and making of chemical bonds*
- *determine enthalpy changes indirectly, using average bond enthalpies*

LEARNING SUMMARY

Bond enthalpy

AQA M2

AQA (A2) M5

Bond enthalpies are **positive** and refer to **bond breaking** – this process requires energy.

Enthalpy is stored within chemical bonds and **bond enthalpy** indicates the strength of a chemical bond in a gaseous molecule. For simple molecules, such as H_2(g) and HCl(g), bond enthalpy applies to the following processes:

$H–H(g) \longrightarrow 2H(g)$ $\Delta H = +436$ kJ mol^{-1}

$H–Cl(g) \longrightarrow H(g) + Cl(g)$ $\Delta H = +432$ kJ mol^{-1}

KEY POINT

Bond enthalpy is the enthalpy change required to **break** and separate **1 mole of bonds** in the molecules of a gaseous element or compound so that the resulting gaseous species exert no forces upon each other.

See also Activation Energy: page 84.

Average bond enthalpies

Not all C–H bonds are created equal.

Bond enthalpies such as those above (H–H and H–Cl) apply to specific compounds. Only H_2 can have H–H bonds and the H–H bond enthalpy shown above has a definite value. However, some bonds can have different strengths in different environments.

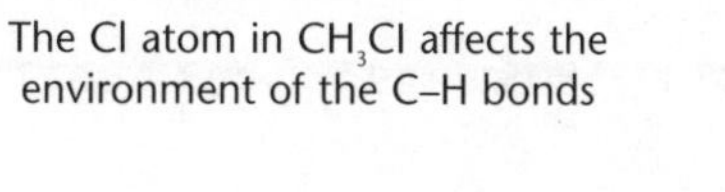

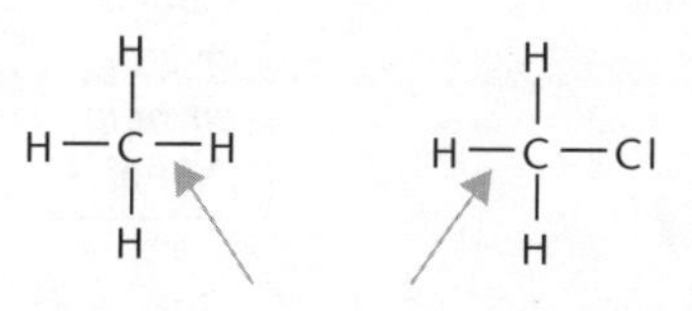

C–H bonds have different strengths and different bond enthalpies

Data books provide an indication of the likely bond enthalpy of a particular bond by listing **average** or **mean bond enthalpies**.

An *average* bond enthalpy indicates the strength of a *typical* bond.
The average bond enthalpy for the C–H bond is +413 kJ mol^{-1}.

Bond making and bond breaking

Bond breaking requires energy: ENDOTHERMIC .

Bond making releases energy: EXOTHERMIC.

Chemical reactions involve bond breaking followed by bond making.

- Energy is first needed to break bonds in the reactants.
  Bond breaking is an endothermic process and **requires** energy.
- Energy is then released as new bonds are formed in the products.
  Bond making is an exothermic process and **releases** energy.

Using bond enthalpies to determine enthalpy changes

Bond enthalpy is an endothermic change (ΔH +ve) for bonds being broken.

When bonds are made, the enthalpy change will be the same magnitude but the opposite sign (ΔH –ve).

The enthalpy change for a reaction involving simple gaseous molecules can be determined using average bond enthalpies in an **energy cycle**:

- Enthalpy required to break bonds = Σ(bond enthalpies in reactants)
- Enthalpy released to make bonds = $-\Sigma$(bond enthalpies in products)

Notice that the relative strengths of the bonds in the reactants and the bonds in the products decide whether a reaction is exothermic or endothermic.

KEY POINT

$\Delta H = \Sigma$(bond enthalpies in reactants) – Σ(bond enthalpies in products).

For the reaction: $CH_4(g) + 2O_2(g) \longrightarrow CO_2(g) + 2H_2O(g)$,

H–C(H)(H)–H + O=O, O=O ⟶ O=C=O + H–O–H, H–O–H

	4 (C–H) + 2 (O=O)	2 (C=O) + 4 (O–H)
ΔH/kJ mol^{-1}	(4 × 413) + (2 × 497)	(2 × 805) + (4 × 463)
	Bonds broken: (endothermic)	*Bonds made: (exothermic)*

$\Delta H = \Sigma$(bond enthalpies in reactants) – Σ(bond enthalpies in products)

$\therefore \Delta H$ = [(4 × 413) + (2 × 497)] – [(2 × 805) + (4 × 463)] kJ mol^{-1}

= **–816 kJ mol^{-1}**

Note that the calculated value is only approximate – the actual bond enthalpies involved may differ from the average values.

- Bonds with small bond enthalpies will break first.
- Low bond enthalpies indicate that a reaction will take place quickly.

Progress check

1 Use the data in the table to find the enthalpy change for each reaction.

(a) $C_2H_4(g) + 3O_2(g) \longrightarrow 2CO_2(g) + 2H_2O(g)$

(b) $N_2(g) + 3H_2(g) \longrightarrow 2NH_3(g)$

bond	C–H	C=O	O=O	O–H	C=C	C–C	N≡N	H–H	N–H
bond average enthalpy /kJmol^{-1}	+413	+805	+497	+463	+612	+347	+945	+436	+391

1 (a) –1317 kJ mol^{-1}; (b) –93 kJ mol^{-1}.

3.4 Reaction rates

After studying this section you should be able to:

- *recall the factors which control the rate of a chemical reaction*
- *understand, in terms of collision theory, the effect of concentration changes on the rate of a reaction*
- *explain the significance of activation energy using the Boltzmann distribution*
- *use the Boltzmann distribution to explain the effect of temperature changes on a reaction rate*
- *understand that an enthalpy change indicates the feasibility of a reaction*

LEARNING SUMMARY

What is a reaction rate?

AQA M2

The rate of a chemical reaction is a measure of how quickly a reaction takes place.

> The rate of a reaction is usually measured as the rate of change of **concentration** of a stated species in a reaction.
> The units of rate are mol dm^{-3} s^{-1}.

KEY POINT

Measuring reaction rates

Although concentration can be measured, other quantities proportional to concentration, such as gas volumes, may be easier to monitor. Rates of different reactions show a very wide variation and it may be more convenient to measure a reaction rate per minute or per hour rather than per second.

What affects a reaction rate?

The reasons why reaction rates are affected by these factors is discussed in the rest of this chapter.

The rate of a reaction may be affected by the following factors:

- the concentration of the reactants
- the surface area of solid reactants
- a temperature change
- the presence of a catalyst.

In some reactions, such as photosynthesis, the rate is affected by the presence and intensity of radiation.

How does reaction rate change during a reaction?

A is being used up – its concentration decreases.

B is being formed – its concentration increases.

For a reaction: **A** ⟶ **B**, the reaction rate = $\frac{\text{change of concentration}}{\text{time}}$

The reaction rate can be measured as:

- the rate of **decrease** in concentration of **A**
- the rate of **increase** in concentration of **B**.

The graphs below show how the concentrations of **A** and **B** change during the course of the reaction: **A** ⟶ **B**.

A is used up and its concentration falls. **B** is formed and its concentration increases.

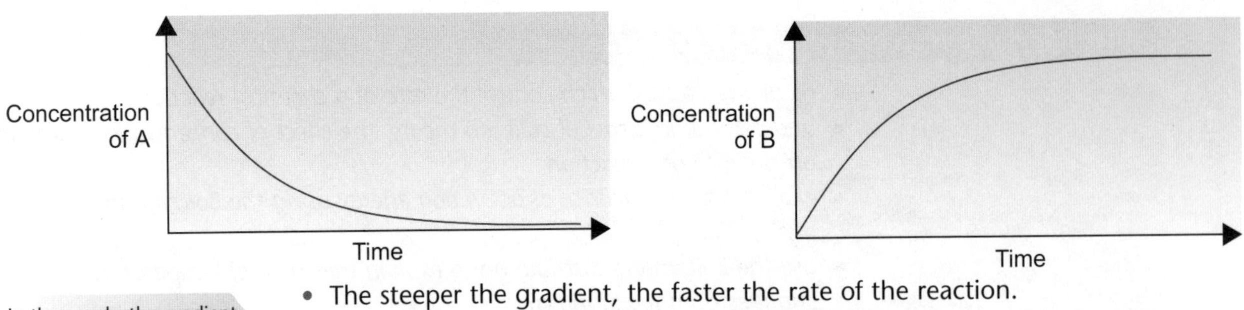

In the graph, the gradient indicates the reaction rate at any time.

- The steeper the gradient, the faster the rate of the reaction.
- The reaction is fastest at the start when the concentration of **A** is greatest.
- As the reaction proceeds, the rate slows down because the concentration of **A** decreases.
- When the reaction is complete, the graph levels off and the gradient becomes zero.

Activation energy

AQA M2

In a gas or solution, particles are in constant motion and they collide with each other, with any solid species and with the walls of their container. When particles collide, a reaction can only take place if the energy of the collision exceeds the activation energy of the reaction.

The activation energy of a reaction is the minimum energy required for the reaction to occur.

Only those collisions of sufficient energy to overcome the activation energy lead to a reaction.

The *activation energy* of a reaction is the energy required to start a reaction by breaking bonds (see page 81).

- Activation energy is often supplied by a spark or by heating the reactants.
- Reactions with a small activation energy often take place very readily.
- A large activation energy may 'protect' the reactants from taking part in the reaction (at room temperature).

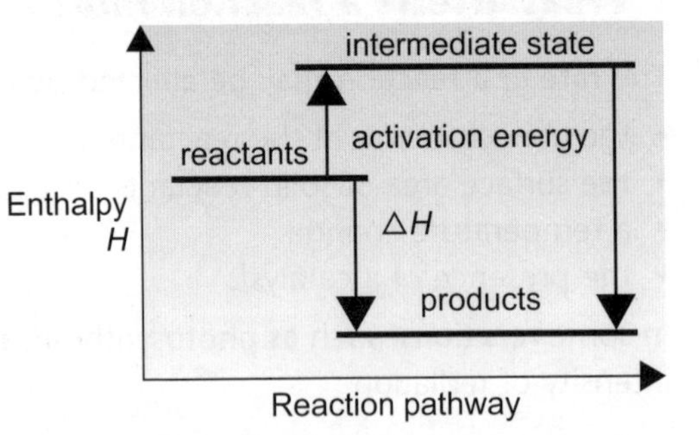

The Boltzmann distribution

AQA M2

The Boltzmann, or Maxwell-Boltzmann, distribution shows the distribution of molecular energies in a gas at constant temperature.

The Boltzmann distribution

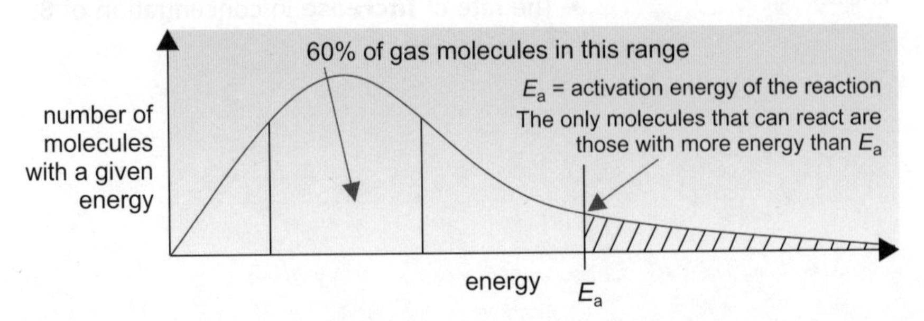

Characteristics of the Boltzmann distribution

- Most gas molecules have energies within a comparatively narrow range.
- The curve will only meet the energy axis at infinity energy. No molecules have zero energy.
- The area under the distribution curve gives the total number of gas molecules.
- Only those molecules with more energy than the activation energy of the reaction are able to react.

The effect of a concentration change on reaction rate

If the concentration of a reactant in solution or in a gas mixture is increased,

- there are more particles present per volume
- more collisions take place each second
- more collisions exceed the activation energy every second
- therefore the rate of reaction increases.

A reaction can only take place if the activation energy is exceeded.

The diagram below shows distribution curves for two concentrations, C_1 and C_2. where concentration, $C_2 >$ concentration, C_1.

- Providing the temperature is the **same**, distribution curves for different concentrations have the **same** shape.

Note that the proportion of the total number of molecules exceeding the activation energy is the same. The rate increases because there are more molecules per volume and more molecules must now exceed the activation energy.

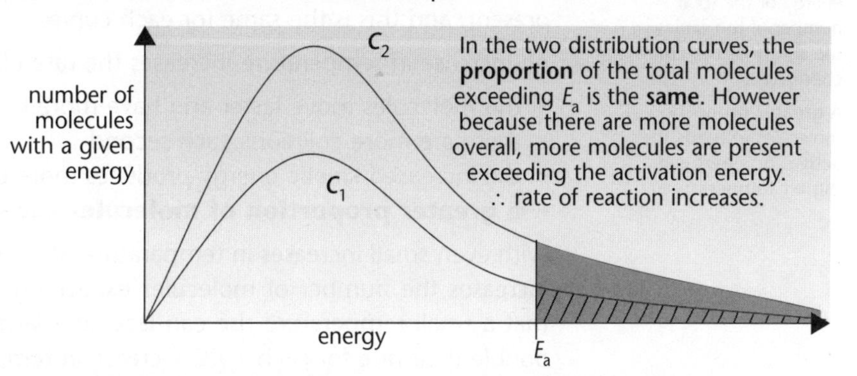

For reactions involving gases, increasing the pressure also increases the concentration of any gas. This results in an increased reaction rate.

The effect of a change in surface area on reaction rate

For a reaction involving a solid, the reaction takes place at a faster rate when the solid is in a powdered form rather than as lumps.

Calcium carbonate reacts with hydrochloric acid producing carbon dioxide gas:

$$CaCO_3(s) + 2HCl(aq) \longrightarrow CaCl_2(aq) + H_2O(l) + CO_2(g)$$

By measuring the volume of carbon dioxide gas evolved with time, the rate of this reaction can be monitored. The graph below compares the reaction rates when using an excess of calcium carbonate as powder and as lumps. The same volume of the same concentration of hydrochloric acid has been reacted in both experiments.

The gradient provides an indication of reaction rate.

The steeper the curve, the faster the reaction.

The final volume is the same in both experiments since the same amount (in mol) of hydrochloric acid has been reacted with an excess of calcium carbonate.

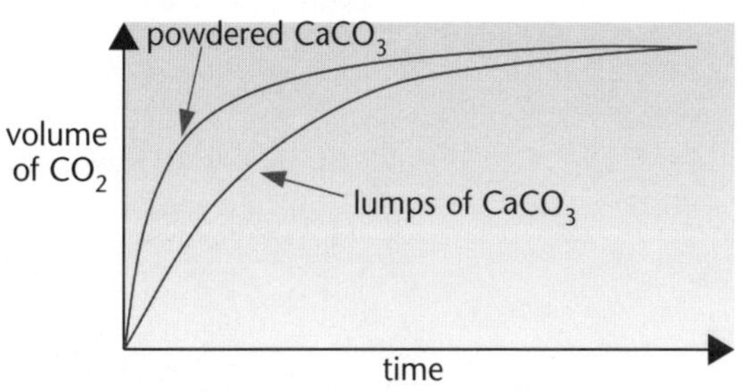

The gradient of the graph is much steeper with powdered carbonate because the surface area available for the reaction with hydrochloric acid is much greater than with lumps, allowing more collisions per second.

The effect of a temperature change on reaction rate

The average kinetic energy of the particles is proportional to temperature. As temperature increases, so does the kinetic energy of gas molecules.

The diagram below shows distribution curves for a sample of gas at two temperatures, T_1 and T_2, where temperature, $T_2 >$ temperature, T_1.

Increasing the temperature **moves** the distribution curve to the right.

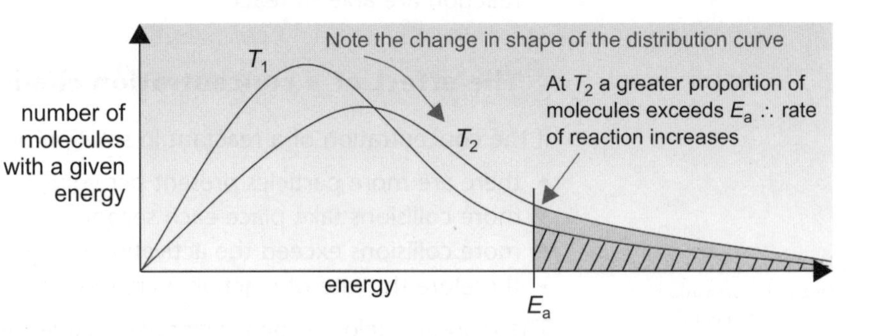

Increasing the temperature **does not change** the activation energy or the total number of molecules – the **shape of the curve changes.**

A **greater proportion** of molecules **exceeds the activation energy at higher temperature.**

The Boltzmann distribution curve is displaced to the right with the peak lower. The average energy is now increased.

The total area under each curve is a measure of the total number of molecules present, and this is the same for each curve.

An increase in temperature increases the rate of a reaction because:

- the molecules move faster and have more kinetic energy
- there are more collisions each second
- the increased kinetic energy produces more energetic collisions
- **a greater proportion of molecules exceed the activation energy**.

With even small increases in temperature, the shift in the distribution curve greatly increases the number of molecules exceeding the activation energy. This means that a small temperature rise can lead to a large increase in rate. Many reactions double their rate for each 10°C increase in temperature.

Progress check

1 Explain the following, in terms of collision theory and distribution graphs.
(a) Reactions take place quicker when the reactants are more concentrated.
(b) Reactions take place quicker at higher temperatures.

1 (a) Because there are more molecules per volume, more collisions take place each second. More collisions exceed the activation energy every second, increasing the reaction rate.
(b) The molecules move faster and have more kinetic energy. There are more collisions each second and the increased kinetic energy produces more energetic collisions. A greater proportion of molecules exceeds the activation energy.

3.5 Catalysis

LEARNING SUMMARY

After studying this section you should be able to:

- *understand what is meant by a catalyst*
- *explain how a catalyst changes the activation energy of a reaction*

How do catalysts work?

AQA M2

What is a catalyst?

A catalyst changes the rate of a chemical reaction, but is unchanged at the end of the reaction. Most catalysts speed up reaction rates although there are some (called inhibitors) that slow down reactions.

These key points are often tested in exams.

KEY POINT

A catalyst speeds up the rate of a reaction by providing an **alternative route** for the reaction with a **lower activation energy**.
Activation energy with catalyst, E_c < Activation energy without catalyst, E_a

The effects of a catalyst on reaction route and activation energy are shown in the diagrams below.

Notice the way the catalyst reduces the energy barrier to the reaction.

The catalyst provides an alternative route in which $E_c < E_a$

The catalyst does **not** change the distribution curve.

Notice how more molecules have an energy exceeding the new, lower activation energy.

Energy profile diagram

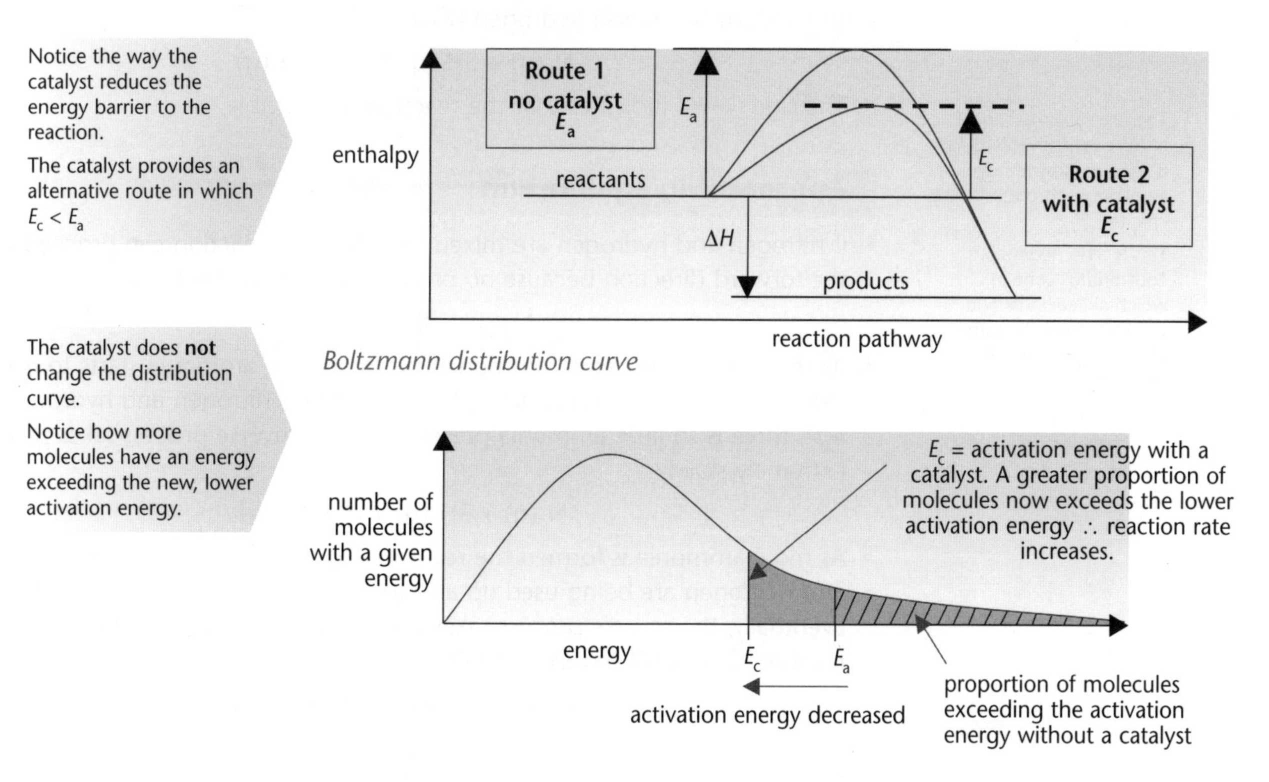

Boltzmann distribution curve

3.6 Chemical equilibrium

After studying this section you should be able to:

- explain the main features of a dynamic equilibrium
- predict the effects of changes on the position of equilibrium
- know that a catalyst does not affect the position of equilibrium
- understand why a compromise temperature and pressure may be used to obtain an economic yield in industrial processes

LEARNING SUMMARY

Reversible reactions and dynamic equilibrium

AQA M2

Reversible reactions

Many chemical reactions take place until the reactants are completely used up and such reactions *'go to completion'*.

In a chemical equation, *reactants* are on the left-hand side; *products* are on the right-hand side.

An example of a reaction that goes to completion is the oxidation of magnesium:

$$2Mg(s) + O_2(g) \longrightarrow 2MgO(s)$$

- The sign '$\longrightarrow$' indicates that the reaction takes place from left to right: this is called the *forward direction*.

Many reactions are **reversible**: they can take place in either direction.

An example of a reaction that is reversible is the formation of ammonia $NH_3(g)$ from nitrogen $N_2(g)$ and hydrogen $H_2(g)$.

$$N_2(g) + 3H_2(g) \rightleftharpoons 2NH_3(g)$$

- The sign '$\rightleftharpoons$' indicates that the reaction is reversible.

The reversible reaction: $N_2(g) + 3H_2(g) \rightleftharpoons 2NH_3(g)$ is used to illustrate dynamic equilibrium throughout this chapter.

This is a homogeneous equilibrium – one in which all reactants and products have the same phase. In this system, all are gases.

Approaching equilibrium

- If nitrogen and hydrogen are mixed together, the reaction can proceed only in the forward direction because no products are yet present:

$$N_2(g) + 3H_2(g) \longrightarrow$$

- As the reaction proceeds, ammonia is formed. The ammonia starts to react in the *reverse direction*, indicated by '$\longleftarrow$', forming nitrogen and hydrogen. At first, there is so little ammonia present that the reverse process takes place extremely slowly:

$$N_2(g) + 3H_2(g) \rightleftarrows 2NH_3(g)$$

- As more ammonia is formed the reverse process takes place faster. Nitrogen and hydrogen are being used up and the forward reaction slows down. Eventually, the reverse process takes place at the same rate as the forward reaction and **equilibrium** is attained:

$$N_2(g) + 3H_2(g) \rightleftarrows 2NH_3(g)$$

Dynamic equilibrium

At equilibrium, there is a balance: reactants and products are both present and the reaction *appears* to have stopped.

- Although there is no apparent change, both forward and reverse processes continue to take place – the equilibrium is *dynamic*.
- The forward reaction proceeds at the same rate as the reverse reaction.

A reversible reaction can be approached from either direction and the terms *reactants* and *products* need to be used with caution.

- The concentrations of reactants and products are constant.

The equilibrium position:

- can be reached from either forward or reverse directions
- can only be achieved in a *closed system* – one in which no materials are being added or removed.

Equilibrium may be approached from either direction

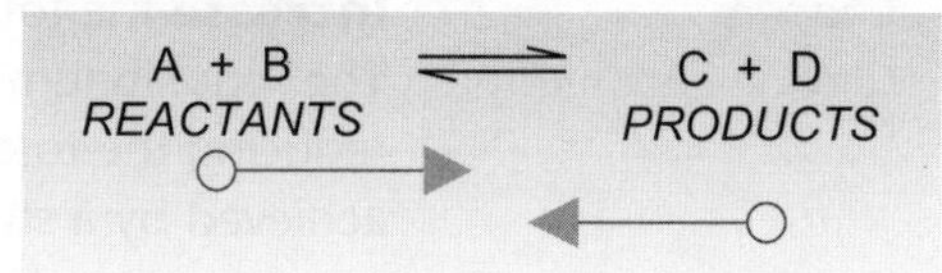

Factors affecting equilibrium

AQA M2

By opening up a closed equilibrium system, conditions can be changed after which the system can be allowed to reach equilibrium again.

The equilibrium position of the system may be altered by the following changes:

- Changing the concentration of a reactant or product.
- Changing the pressure of a gaseous equilibrium.
- Changing the temperature.

The likely effect on the equilibrium position can be predicted using Le Chatelier's Principle.

> *Le Chatelier's Principle* states that if a system in dynamic equilibrium is subjected to a change, processes will occur to minimise this change.
>
> KEY POINT

The effect of concentration changes

A change in the concentration of a reactant will alter the rate of the forward direction. A change in the concentration of a product will alter the rate of the reverse direction. Either change results in a shift in the equilibrium position.

$$N_2(g) + 3H_2(g) \rightleftharpoons 2NH_3(g)$$

Change: **Increase** concentration of a reactant: $N_2(g)$ or $H_2(g)$

Change opposed: The equilibrium system will **decrease** the concentration of the reactant by removing it. This is achieved by a shift in equilibrium to the right, forming more $NH_3(g)$.

If the concentration of a reactant is increased, or the concentration of the product is decreased, e.g. by removing some of it, the equilibrium is displaced to the right and more product is obtained.

The effects of changes in concentration for this equilibrium are shown below. These effects will apply to any system in equilibrium.

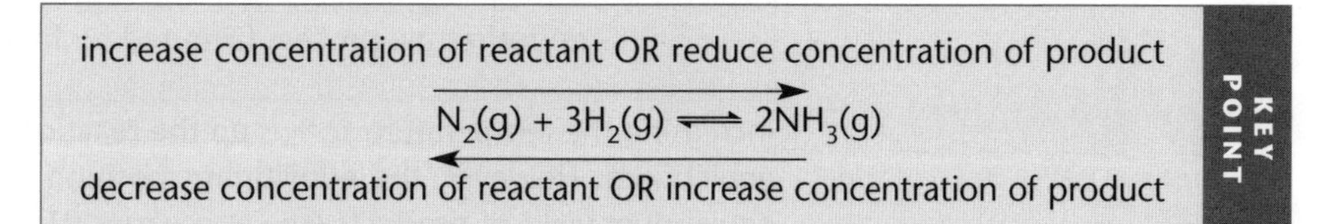

KEY POINT

The effect of pressure changes

A change in total pressure may alter the equilibrium position of a system involving gases. **The direction favoured depends upon the total number of gas molecules on each side of the equilibrium.**

$$N_2(g) + 3H_2(g) \rightleftharpoons 2NH_3(g)$$

Change: **Increase** the total pressure.

Change opposed: The equilibrium system will **decrease** the pressure by reducing the total number of moles of gas molecules. This is achieved by a shift in equilibrium to the right:

$N_2(g)$	+	$3H_2(g)$	$\rightleftharpoons$	$2NH_3(g)$
1 mol		3 mol		2 mol
	4 mol			2 mol

Movement of gas molecules causes pressure. The more gas molecules in a volume, the greater the pressure.

If there are more moles of gaseous reactant than there are moles of gaseous product, an increase in total pressure will displace the reaction to the right.

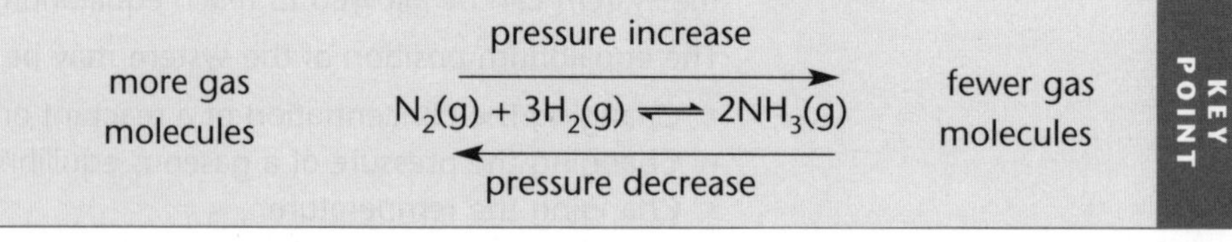

Increasing the pressure also increases the concentration of any gas present and this **speeds** up the reaction.

The equilibrium position is also changed if:

- at least one of the equilibrium species is a gas
- there are different numbers of gaseous moles of reactants and of products.

An exothermic process is favoured by low temperatures.

An endothermic process is favoured by high temperatures.

The effect of temperature changes

A change in temperature alters the rates of the forward and reverse reactions by different amounts, resulting in a shift in the equilibrium position. **The direction favoured depends upon the sign of the enthalpy change.**

$$N_2(g) + 3H_2(g) \rightleftharpoons 2NH_3(g) \qquad \Delta H^\ominus = -92 \text{ kJ mol}^{-1}$$

Change: **Decrease** the temperature.

Change opposed: The equilibrium system will **increase** the temperature by releasing more heat energy. This is achieved by a shift in equilibrium in the exothermic direction, to the right.

An exothermic reaction in one direction is endothermic in the opposite direction.

The value of the enthalpy changes is the same – the sign is different.

KEY POINT

exothermic process favoured by low temperatures

$$\Delta H^\ominus = +92 \text{ kJ mol}^{-1} \quad N_2(g) + 3H_2(g) \rightleftharpoons 2NH_3(g) \quad \Delta H^\ominus = -92 \text{ kJ mol}^{-1}$$

endothermic process favoured by high temperatures

Increasing the temperature *speeds* up the reaction and less time is needed to reach equilibrium. However, the equilibrium position may change to produce a lower equilibrium yield of products (see also pages 91–92).

The effect of a catalyst

There is **no change** in the equilibrium position. However, the catalyst **speeds** up both forward and reverse reactions and equilibrium is reached quicker.

Equilibria, rates and industrial processes

AQA M2 OCR M2
CCEA M2

For many industrial processes, the overall yield of a chemical product can vary considerably, depending on the conditions used.

In deciding the conditions to use, industrial chemists need to consider:

- the availability of the starting materials required for a process
- the equilibrium conditions required to ensure a good yield
- the rate of reaction, which should be fast but manageable
- the cost, which should be as low as possible taking into account energy, cost of materials and of the chemical plant
- the safety of workers from hazardous chemicals, pressure, temperature, etc.
- any effects on the environment from waste discharges, toxic fumes, etc.

All courses study one or more industrial processes.

You should learn outline details of any process on your course.

The principles, however, are common for any similar situation.

Industrial chemists compare each of these factors to arrive at compromise conditions. Although the equilibrium yield might be **optimised** by using a very high temperature and pressure, this may prove **impracticable** for reasons of cost and safety.

The importance of compromise in industrial processes

AQA M2

Four industrial processes are discussed below to illustrate the importance of compromise.

In each of the four processes:

- the right-hand side, with the desired product, has **fewer moles of gas** than the left-hand side (reactants)
- the forward reaction, forming the desired product, is **exothermic**.

The raw materials used should be readily available.

$N_2(g)$ is obtained from the air.

$H_2(g)$ is obtained, together with $CO(g)$, by reacting together natural gas, $CH_4(g)$, and steam, $H_2O(g)$.

$C_2H_4(g)$ is obtained from the cracking of crude oil fractions.

The Haber process for the production of ammonia

$$N_2(g) + 3H_2(g) \rightleftharpoons 2NH_3(g) \qquad \Delta H^\ominus = -92 \text{ kJ mol}^{-1}$$

The contact process during the production of sulfuric acid

$$2SO_2(g) + O_2(g) \rightleftharpoons 2SO_3(g) \qquad \Delta H^\ominus = -197 \text{ kJ mol}^{-1}$$

The industrial production of ethanol by hydration of ethene

$$C_2H_4(g) + H_2O(g) \rightleftharpoons C_2H_5OH(g) \qquad \Delta H^\ominus = -46 \text{ kJ mol}^{-1}$$

The industrial production of methanol from carbon monoxide with hydrogen

$$CO(g) + 2H_2(g) \rightleftharpoons CH_3OH(g) \qquad \Delta H^\ominus = -91 \text{ kJ mol}^{-1}$$

The optimum equilibrium conditions

Optimising the process

These are the ideal conditions to give a maximum equilibrium yield.

Consider le Chatelier's Principle:

- The forward reaction produces fewer moles of gas, favoured by a **high** pressure.
- The forward reaction producing the desired product is exothermic, favoured by a low temperature.

The optimum equilibrium conditions for maximum equilibrium yield are:

- **high pressure and low temperature.**

Compromising

Optimum conditions, reaction rate, feasibility, reality, safety and economics must be considered.

The need for compromise

Having arrived at optimum equilibrium conditions, each must be considered to arrive at compromise conditions.

optimum condition	*advantages*	*disadvantages*
HIGH PRESSURE	equilibrium yield of desired product is high	energy costs are high – it is expensive to compress gases
	the concentration of gases is high, increasing the rate	there are considerable safety implications of using very high pressures – vessel walls need to be very thick to withstand high pressures, and weaknesses cause danger to workers and potential leakage of chemicals into the environment
LOW TEMPERATURE	equilibrium yield of product is high	the reaction takes place very slowly as few molecules possess the activation energy of the reaction

The use of a catalyst

A catalyst speeds up the rates of both the forward and backward reactions and hence less time is needed for the reaction to reach equilibrium. The increase in reaction rate allows lower temperatures to be used for a realistic reaction rate. The use of lower temperatures has the added bonus of supporting the optimum conditions for this process, increasing the equilibrium yield of the desired product.

Exam papers always test the wider aspects of chemistry to society.

Exam answers need to be sensible and related to relevant issues arising from the process itself. It is pointless to use terms such as 'dangerous', 'toxic', etc. unless substantiated with the reasons why the hazard is present.

Compromise conditions

The compromise conditions used in each process are such that:

- the temperature is sufficiently high to allow the reaction to occur at a realistic rate, but not too high to give a minimal equilibrium yield of the desired product
- a high pressure is used but not too high as to be impracticable
- the reaction rate is increased by using a catalyst. This enables the process to take place at lower temperatures, saving on energy costs.

The compromise conditions used for a process may only yield a small percentage of the desired product. Unreacted reactants are recycled.

Progress check

1 What will be the result of an increase in pressure on the following equilibria?

(a) $N_2O_4(g) \rightleftharpoons 2NO_2(g)$

(b) $CO(g) + 2H_2(g) \rightleftharpoons CH_3OH(g)$

(c) $H_2(g) + Br_2(g) \rightleftharpoons 2HBr(g)$

2 What will be the result of an increase in temperature on the following equilibria?

(a) $N_2(g) + O_2(g) \rightleftharpoons 2NO(g)$ $\Delta H^{\ominus} = +180$ kJ mol^{-1}

(b) $H_2(g) + Br_2(g) \rightleftharpoons 2HBr(g)$ $\Delta H^{\ominus} = -9.6$ kJ mol^{-1}

(c) $2SO_2(g) + O_2(g) \rightleftharpoons 2SO_3(g)$ $\Delta H^{\ominus} = -197$ kJ mol^{-1}

3 What are the optimum conditions of temperature **and** pressure for a high yield of the products in the equilibria below?

(a) $CO(g) + 2H_2(g) \rightleftharpoons CH_3OH(g)$ $\Delta H^{\ominus} = -92$ kJ mol^{-1}

(b) $PCl_5(g) \rightleftharpoons PCl_3(g) + Cl_2(g)$ $\Delta H^{\ominus} = +124$ kJ mol^{-1}

1 (a) moves to left; (b) moves to right; (c) no change.

2 (a) moves to right; (b) moves to left; (c) moves to left.

3 (a) low temperature and high pressure; (b) high temperature and low pressure.

Sample question and model answer

1

The diagram below shows the Boltzmann distribution at a temperature, T_1, for a mixture of gases that react with each other. The activation energy for the reaction is labelled **X**.

Notice that the distribution curve starts at the origin but it never quite reaches the x-axis, even at high energies. Remember also that the area under the curve is equal to the total number of molecules.

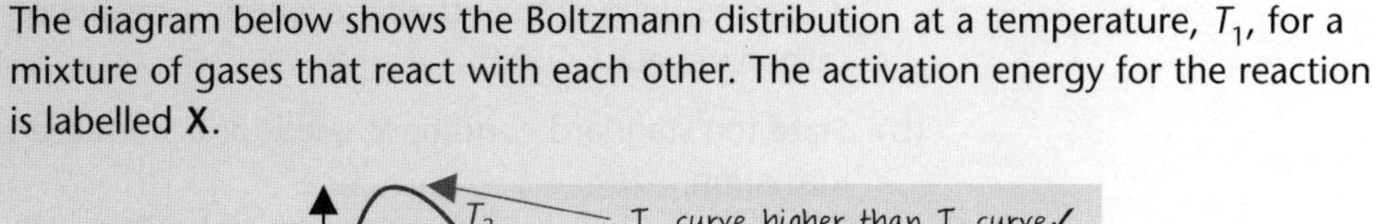

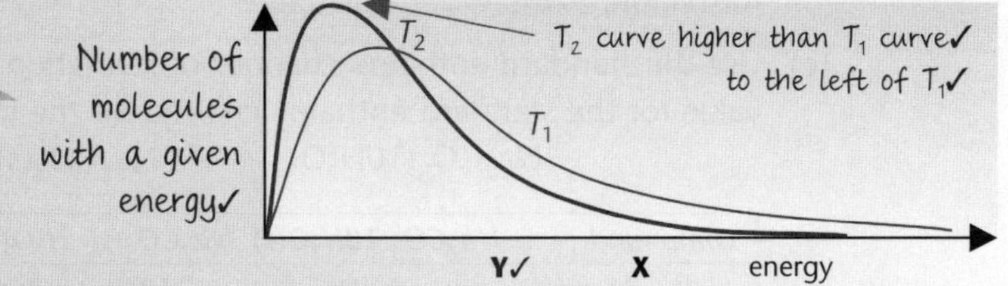

(a) (i) Explain the meaning of the term *activation energy*.

This is the minimum ✓ energy required for a reaction to take place. ✓

(ii) Label the y-axis.

(iii) Draw a second curve on the diagram above for the same mixture at a lower temperature. Label the second curve T_2.

(iv) Explain how a lower temperature affects the rate of this reaction.

Reduced temperature slows down the reaction rate. ✓
The overall kinetic energy of the molecules is reduced ✓ and fewer molecules now exceed the activation energy. ✓ [8]

At different temperatures,
the curve changes
the activation energy stays the same.

Note the difference

In the presence of a catalyst,
the activation energy changes
the curve stays the same.

(b) A catalyst is added to the mixture of gases.

(i) On the diagram above, label a possible activation energy **Y** for the catalysed reaction.

(ii) Explain, in terms of activation energy and the Boltzmann distribution, how a catalyst affects the rate of this reaction.

The catalyst speeds up the reaction rate. ✓ A catalyst allows the reaction to proceed via a different route ✓ with a lower activation energy. This means that a larger proportion of molecules now exceeds the activation energy. ✓ [4]

(c) The equation for the industrial manufacture of ethanol, C_2H_5OH, from ethene, C_2H_4, is shown below.

$$C_2H_4(g) + H_2O(g) \rightleftharpoons C_2H_5OH(g) \qquad \Delta H = -46\ \text{kJ mol}^{-1}$$

The industrial conditions used are a high temperature and high pressure.

Always pay attention to the words that examiners use.

In parts (i) and (ii), the examiner has used 'Explain'. You need to justify your answer.

In part (iii), the examiner has used 'State'. You only need to 'state' your answer. You waste time here if you 'explain'.

(i) Explain why the reaction is carried out at a high pressure.
There is a greater equilibrium yield. ✓ The increase in pressure is relieved by reducing the number of gas molecules. ✓ The equilibrium shifts in favour to the right because there are fewer gas molecules to the right. ✓

(ii) Explain why pressures higher than 1000 atmospheres are not used.
High pressures are very costly to generate in terms of energy. ✓

(iii) State **one** advantage and **one** disadvantage of carrying out the reaction at high temperature.
An advantage is that there is a greater reaction rate. ✓
A disadvantage is that the equilibrium yield is reduced. ✓ [6]

[Total: 18]

Practice examination questions

1 (a) Write a chemical equation, including state symbols, for the reaction that is used to define the enthalpy change of formation of one mole of sodium carbonate, $Na_2CO_3(s)$. [2]

(b) State the standard conditions used for a standard enthalpy change of formation, $\Delta H^{\ominus}_f$. [1]

(c) Use the standard enthalpy changes of formation given below to calculate a value for the standard enthalpy change for the following reaction:

$$Na_2CO_3.10H_2O(s) \longrightarrow Na_2CO_3(s) + 10H_2O(l)$$

compound	$Na_2CO_3.10H_2O(s)$	$Na_2CO_3(s)$	$H_2O(l)$
$\Delta H^{\ominus}_f$ / kJ mol^{-1}	–4081	–1131	–286

[3]

[Total: 6]

2 Propan-1-ol, $CH_3CH_2CH_2OH$, reacts with oxygen in a combustion reaction.

(a) (i) Define the term *standard enthalpy change of combustion.*

(ii) State the temperature that is conventionally chosen for standard enthalpy changes. [4]

(b) (i) Write a balanced equation for the combustion of propan-1-ol.

(ii) Calculate the standard enthalpy change of combustion of propan-1-ol using the following data.

compound	$\Delta H^{\ominus}_f$ /kJ mol^{-1}
$CH_3CH_2CH_2OH(l)$	–303
$CO_2(g)$	–394
$H_2O(l)$	–286

[4]

[Total: 8]

3 (a) Define the term *standard enthalpy change of formation* ($\Delta H^{\ominus}_f$). [3]

(b) Give the equation for the change that represents the standard enthalpy change of formation of methane. [2]

(c) Use the following data to calculate a value for the standard enthalpy change of formation of methane.

compound	$\Delta H^{\ominus}_c$ /kJ mol^{-1}
C(s)	–394
$H_2(g)$	–242
$CH_4(g)$	–802

[3]

(d) Use the data below to calculate average bond enthalpy values for the C–H and the C–C bonds.

$CH_4(g) \longrightarrow C(g) + 4H(g)$ $\quad \Delta H = 1648$ kJ mol^{-1}

$C_2H_6(g) \longrightarrow 2C(g) + 6H(g)$ $\quad \Delta H = 2820$ kJ mol^{-1} [3]

[Total: 11]

Practice examination questions

4 (a) The diagram below shows a Boltzmann distribution for the mixture of gases at a temperature, T_1.

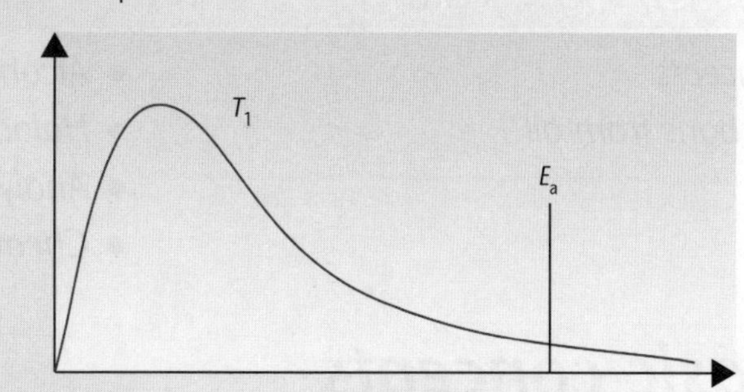

(i) Label the axes and add a second curve for the Boltzmann distribution of this mixture at a higher temperature, T_2.

(ii) On the diagram, add a label: 'E_a catalysed' for a possible activation energy of the reaction when catalysed. [5]

(b) The elimination of steam from ethanol is an endothermic reaction:

$$C_2H_5OH(g) \longrightarrow C_2H_4(g) + H_2O(g) \qquad \Delta H = +46 \text{ kJ mol}^{-1}$$

(i) Sketch the reaction profile diagram for this reaction. On your diagram, label clearly the enthalpy change for the reaction, ΔH, and the activation energy, E_a.

(ii) Add to your diagram a labelled reaction profile diagram for this reaction when catalysed.

(iii) Explain why a catalyst increases the rate of a chemical reaction.

[7]

[Total: 12]

5 When hydrogen and carbon dioxide react, a dynamic homogenous equilibrium is set up as shown below:

$$H_2(g) + CO_2(g) \rightleftharpoons H_2O(g) + CO(g)$$

(a) (i) Why is this reaction a *homogeneous equilibrium*?

(ii) State **three** features of a *dynamic equilibrium*. [5]

(b) For this equilibrium, explain the effect of an increase in pressure on:

(i) the equilibrium position

(ii) the reaction rate. [4]

(c) The equilibrium yield of steam and carbon monoxide increases as the temperature is increased. Determine whether the forward reaction in the equilibrium above is exothermic or endothermic. Explain your answer. [2]

[Total 11]

Chapter 4

Organic chemistry, analysis and the environment

The following topics are covered in this chapter:

- Basic concepts
- Hydrocarbons from oil
- Alkanes
- Alkenes
- Alcohols
- Haloalkanes
- Analysis
- Chemistry in the environment

4.1 Basic concepts

After studying this section you should be able to:

- understand the different types of formula used for organic compounds
- recognise types of hydrocarbon
- recognise common functional groups
- understand what is meant by structural isomerism
- apply rules for naming simple organic compounds
- understand the difference between homolytic and heterolytic fission
- calculate percentage yields and atom economies

LEARNING SUMMARY

Types of formula

AQA M1

In organic chemistry, there are many ways of representing a formula.

For the compound butane, with 4 carbon atoms and 10 hydrogen atoms:

the *empirical* formula is:	C_2H_5	The simplest, whole-number ratio of elements in a compound.
the *molecular* formula is:	C_4H_{10}	The *actual* number of atoms of each element in a molecule.
the *structural* formula is:	$CH_3CH_2CH_2CH_3$	The minimal detail for an unambiguous structure.
the *displayed* formula is:	H–C–C–C–C–H (each C bonded to H above and H below)	The relative placing of atoms and the bonds between them.
the *skeletal* formula		The carbon skeleton and functional groups only.

Carbon chains

AQA M1

Hydrocarbons

Hydrocarbons are compounds of carbon and hydrogen **only**.

- A saturated hydrocarbon has single bonds only.
- An unsaturated hydrocarbon contains a multiple carbon carbon bond.

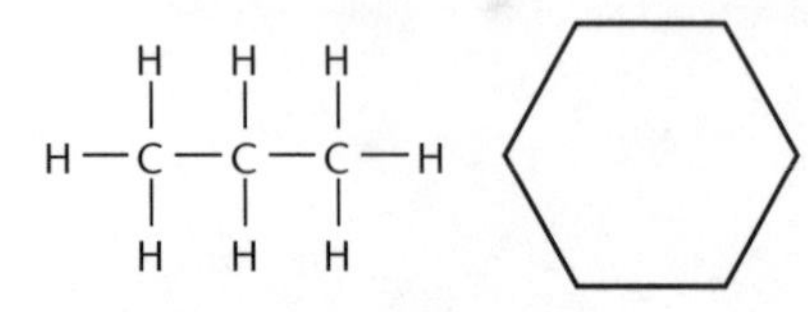

saturated – single bonds only

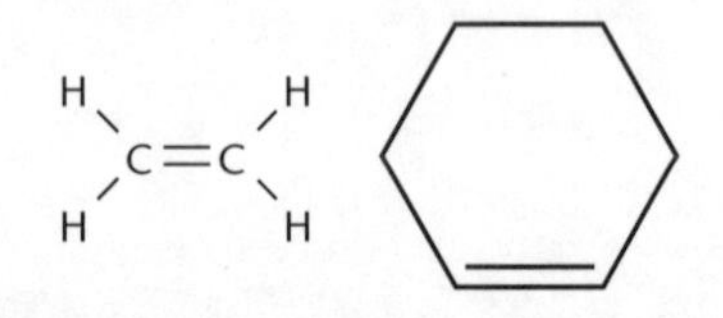

unsaturated – contains a double bond

Alkanes

Carbon atoms can bond with other carbon atoms to form an enormous range of compounds with different carbon-chain lengths. The simplest organic compounds are a family of saturated hydrocarbons called the alkanes, shown below.

meth- C
eth- C_2
prop- C_3
but- C_4
pent- C_5
hex- C_6
hept- C_7
oct- C_8
non- C_9
dec- C_{10}

Number of carbons	*Name*	*Molecular formula*	*Structural formula*
1	methane	CH_4	CH_4
2	ethane	C_2H_6	CH_3CH_3
3	propane	C_3H_8	$CH_3CH_2CH_3$
4	butane	C_4H_{10}	$CH_3CH_2CH_2CH_3$
5	pentane	C_5H_{12}	$CH_3CH_2CH_2CH_2CH_3$
6	hexane	C_6H_{14}	$CH_3CH_2CH_2CH_2CH_2CH_3$

Note the following points.

- The name of the alkane ends with *-ane.*
- The prefixes (*meth-*, *eth-*, ...) are used to represent the number of carbon atoms. You will need to use these many times in organic chemistry and they must be learnt.

General formula

Alkanes, C_nH_{2n+2} are saturated.

- The **general formula** is the simplest algebraic representation for any member in a series of organic compounds.
- The general formula for any alkane is C_nH_{2n+2}, where n = the number of carbon atoms.

Homologous series

Members in a homologous series react similarly.
By studying the reactions of one member of the series, you know how all members in the series are likely to react.

The alkanes are an example of a **homologous series** with the following features.

- Each successive member differs by $-CH_2-$. You can see this by comparing the formula of each alkane in the table above.
- Each member in a homologous series has the same general formula. The alkanes have the general formula: C_nH_{2n+2}.
- All members in a homologous series have the same functional group and similar chemical reactions.

Note that physical properties, such as boiling point and density, do gradually change as the length of the carbon chain increases.

Functional groups

AQA M1

Saturated hydrocarbon chains are comparatively unreactive. The reactivity is increased by the presence of a functional group – the reactive part of a carbon compound.

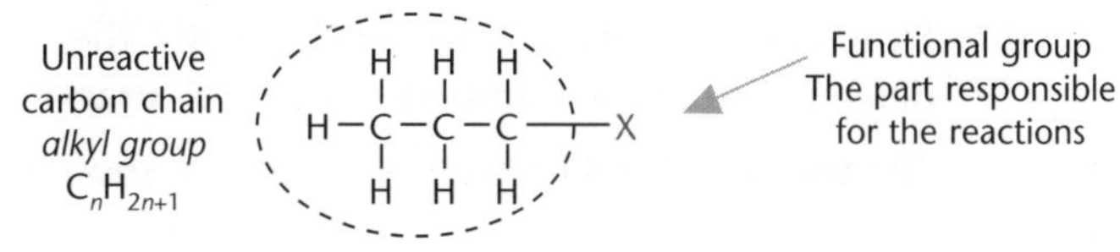

Common functional groups

It is essential that you can instantly identify a functional group within a molecule. You can then apply the relevant chemistry.

If you continue to study Chemistry to A Level, you will meet more functional groups.

The main functional groups for AS Chemistry are shown below.

name	*functional group*	*structural formula*	*general formula*	*prefix or suffix (for naming)*
alkane	C–H	$CH_3CH_2CH_3$ *propane*	C_nH_{2n+2}	-ane
alkene	>C=C<	$CH_3CH{=}CH_2$ *propene*	C_nH_{2n}	-ene
halogenoalkane	C–X (X=halogen)	CH_3CH_2Br *bromoethane*	$C_nH_{2n+1}Br$	bromo-
alcohol	C–OH	CH_3CH_2OH *ethanol*	$C_nH_{2n+1}OH$	-ol
aldehyde	–C(=O)H	CH_3CHO *ethanal*	$C_nH_{2n}O$	-al
ketone	R–C(=O)–R	$CH_3COC_2H_5$ *butanone*	$C_nH_{2n}O$	-one
carboxylic acid	–C(=O)OH	CH_3COOH *ethanoic acid*	$C_nH_{2n+1}COOH$	-oic acid

- When studying reactions of a homologous series, the alkyl group is relatively unimportant – it is the functional group that reacts.
- The alkyl group, C_nH_{2n+1} is often represented simply as R–. This shifts the emphasis in the formula towards the reactive part of the molecule, the functional group.

The alcohols, shown below, is a homologous series with the –OH functional group.

- The alcohols have the general formula: $C_nH_{2n+1}OH$
- *Any* alcohol can be represented as: R–OH

The alcohols

CH_3OH → (CH_2) → C_2H_5OH → (CH_2) → C_3H_7OH

Same functional group, OH

Structural isomerism

AQA M1

In addition to forming long chains, the atoms making up a molecular formula are often arranged differently, forming **isomers**.

Structural isomerism

Structural isomers are molecules with the same molecular formula but with different structural formulae (structural arrangements of atoms).

The two structural isomers of C_4H_{10} are shown below:

See also *E/Z* isomerism, page 106.

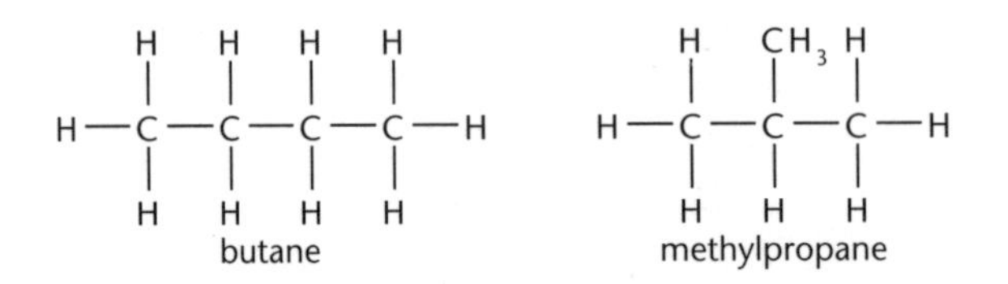

Butane and methylpropane are chain isomers of C_4H_{10}. Chain isomerism is a type of structural isomerism in which the carbon skeleton is different.

The number of different structural isomers possible from a molecular formula increases dramatically as the carbon-chain length increases. This is shown in the table below.

molecular formula	*number of isomers*
C_5H_{12}	3
C_6H_{14}	5
C_7H_{16}	9
C_8H_{18}	18
C_9H_{20}	35
$C_{10}H_{22}$	75
$C_{15}H_{32}$	4,347
$C_{20}H_{42}$	366,319

Look at the different names given to the two isomers.

- The branched isomer, methylpropane, is treated as a side chain attached to a straight-chain alkane.
- Side chains, such as the methyl group CH_3–, are called *alkyl* groups with the general formula of C_nH_{2n+1}.
- An alkyl group can be regarded as an alkane with one hydrogen atom removed (to allow another group to be attached).

The naming of organic compounds is discussed in more detail below.

Notice how the alkyl groups are named and how their formulae are linked to the parent alkane.

You must learn these.

Alkanes and alkyl groups

Number of carbons	*Alkane C_nH_{2n+2} Formula*	*Name*	*Alkyl group C_nH_{2n+1} Formula*	*Name*
1	CH_4	methane	CH_3–	methyl
2	C_2H_6	ethane	C_2H_5–	ethyl
3	C_3H_8	propane	C_3H_7–	propyl

In position isomerism, a functional group can be at different positions on the chain. For example, there are two position isomers with the molecular formula $C_4H_{10}O$ that have the alcohol, –OH, functional group:

- butan-1-ol with the –OH functional group at the end of the carbon chain
- butan-2-ol with the –OH functional group one carbon in from the end of the carbon chain.

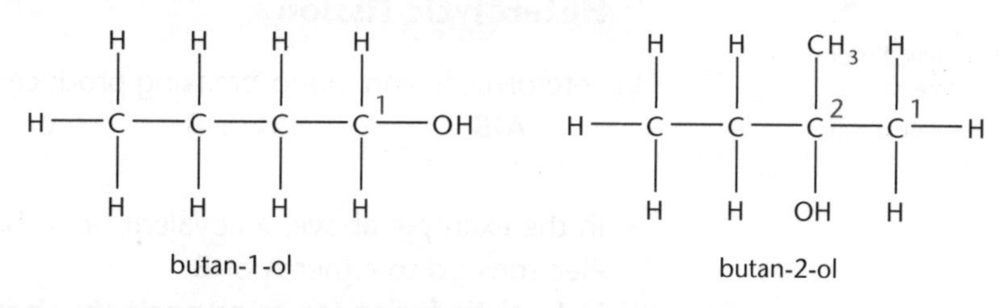

Naming of organic compounds

AQA M1

With so many isomers possible, it is important that each has an individual name. The examples below show the steps needed to name an organic compound.

Rule 1

- The name is based upon the longest carbon chain on an **alkane**.

Rule 2

- Any functional groups and alkyl groups are identified.
- These are then added to the name as a prefix (e.g. chloro-) or suffix (e.g. -ol) (see page 98).

Rule 3

- If there is more than one possible isomer then the carbon atoms are labelled with numbers. Numbering starts from the end giving the lowest possible numbers for any functional groups and side chains. In this example, numbering from the left would give the incorrect name of 3-methylbutan-4-ol.

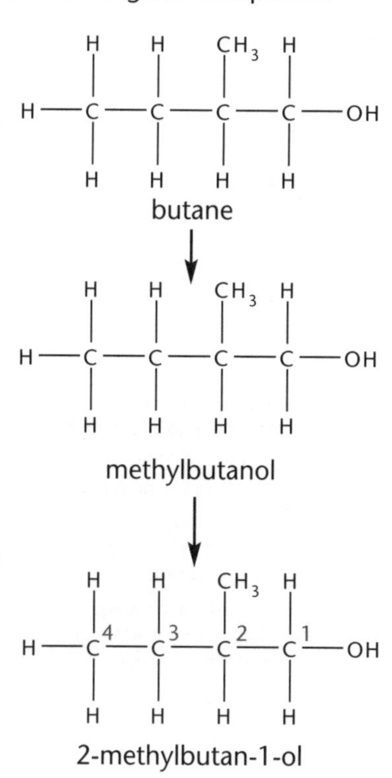

Rule 4

- If there is more than one alkyl or functional group, they are placed in alphabetical order. This rule is not needed for the example above but it does need to be applied to name the compound to the right.

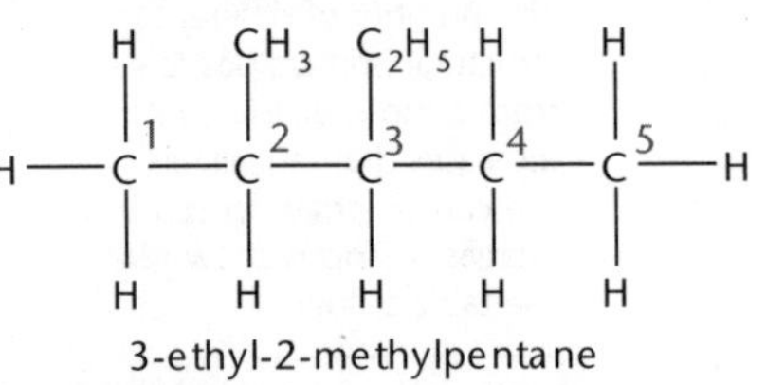

3-ethyl-2-methylpentane

Types of bond fission

AQA M2

The breaking of a covalent bond is called **bond fission**. Reactions of organic compounds involve bond fission followed by the formation of new bonds. Two types of bond fission are possible, **homolytic** fission and **heterolytic** fission.

Homolytic fission

These are very important principles, used throughout organic chemistry.

In homolytic fission, bond-breaking produces two species of the same (*homo-*) type:

A:B ⟶ **A• + •B**
free radicals

A free radical is a species with an unpaired electron (see page 104–105).

- In the example above, a covalent bond breaks so that one of the bonding electrons goes to each of **A** and **B**.
- Homolytic fission forms **two free radicals**.

Heterolytic fission

Homolytic fission ⟶ free radicals

Heterolytic fission ⟶ ions

In heterolytic fission, bond-breaking produces two species of different (*hetero-*) type:

A:B ⟶ **A:⁻ + B⁺** *ions* or **A:B** ⟶ **A⁺ + :B⁻** *ions*

- In the example above, a covalent bond breaks so that both the bonding electrons go to either **A** or **B**.
- Heterolytic fission forms **oppositely-charged ions**.

Percentage yields

AQA M1

Organic reactions typically produce a lower yield than expected from the balanced equation. The yield is usually expressed as a percentage yield:

$$\text{percentage yield} = \frac{\text{actual yield}}{\text{theoretical yield}} \times 100\%$$

KEY POINT

Example

See also page 31.

5.2 g of 1-bromobutane is formed by reacting 5.0 g of butan-1-ol with sodium bromide and concentrated sulfuric acid. Find the percentage yield of 1-bromobutane.

equation	$C_4H_9OH + NaBr + H_2SO_4$	⟶	$C_4H_9Br + H_2O + NaHSO_4$
moles	1 mol	⟶	1 mol
reacting masses	74.0 g	⟶	136.9 g
	1 g	⟶	$\frac{136.9}{74.0}$ g
	5.0 g	⟶	$5 \times \frac{136.9}{74.0}$ g

Factors that can contribute to a low yield:

Organic reactions often do not go to completion.

An organic compound often reacts to produce a mixture of products.

The purification stages result in loss of some of the desired product.

∴ 5.0 g of C_4H_9OH produces a theoretical yield of 9.3 g C_4H_9Br.

Mass of C_4H_9Br that forms = 5.2 g.

$$\text{percentage yield} = \frac{\text{actual yield}}{\text{theoretical yield}} \times 100\% = \frac{5.2}{9.3} \times 100 = 56\%$$

Atom economy

AQA M1

We are all becoming far more aware of our environment and the need to conserve resources and produce less waste. Atom economy is used to assess the efficiency of a reaction in terms of atoms:

KEY POINT

$$\text{atom economy} = \frac{\text{molecular mass of the desired product}}{\text{sum of molecular masses of all products}} \times 100$$

Atom economy is calculated from the balanced equation for the reaction.

> To improve the atom economy for C_4H_9Br production, a chemical company must either find uses for the other products or use another method for C_4H_9Br production.

Example

equation: $C_4H_9OH + NaBr + H_2SO_4 \longrightarrow C_4H_9Br + H_2O + NaHSO_4$

molecular mass of desired product:

$C_4H_9Br = 136.9$

sum of molecular masses of all products:

$C_4H_9Br + H_2O + NaHSO_4 = 136.9 + 18.0 + 120.1 = 275.0$

$$\text{atom economy} = \frac{136.9}{275.0} \times 100 = 49.8\%$$

The consequence is that, even if this reaction could be carried out with a 100% percentage yield, converting **all** of the organic starting material, C_4H_9OH, into the organic product, C_4H_9Br, more than half the mass of products is **waste**.

> An addition reaction produces a single product: all atoms are used and the reaction has an atom economy of 100%.

> Substitution and elimination always produce by-products.
> These are wasted atoms, unless the by-products can be used.

Atom economy and type of reaction

Addition reactions are the most atom efficient, giving a single product and an atom economy of 100%.

addition: $CH_2{=}CH_2 + H_2O \longrightarrow C_2H_5OH$

Substitution and elimination reactions produce by-products and will always have less atom economy unless some use is found for by-products.

substitution: $CH_3CH_2CH_2Br + NaOH \longrightarrow CH_3CH_2CH_2OH + NaBr$

elimination: $CH_3CH_2CH_2Br \longrightarrow CH_3CH{=}CH_2 + HBr$

Progress check

1. C_6H_{14} has 5 structural isomers. Show the structural and displayed formula for each of these isomers and name them.
2. 8.52 g of 1-chloropentane was reacted with aqueous sodium hydroxide, producing 4.75 g of pentan-1-ol.
   $CH_3CH_2CH_2CH_2CH_2Cl + NaOH \longrightarrow CH_3CH_2CH_2CH_2CH_2OH + NaCl$
   Find the percentage yield of 1-bromobutane and the atom economy of the reaction.

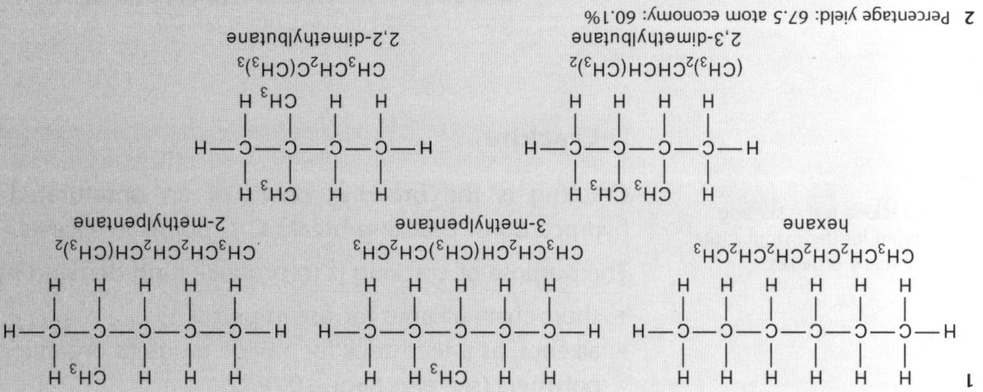

4.2 Hydrocarbons from oil

After studying this section you should be able to:

- *understand how hydrocarbon fractions are separated from crude oil*
- *describe how low demand fractions are processed into hydrocarbons of higher demand*

LEARNING SUMMARY

Useful products from crude oil

AQA M1

Crude oil is a fossil fuel, formed from the decay of sea creatures over millions of years. It is a complex mixture of hydrocarbons, containing mainly alkanes. Crude oil cannot be used directly but it is the source of many useful products. The first stages in the processing of crude oil are described below.

Fractional distillation

Crude oil is heated and passed into a fractionating tower, separating the complex mixture into *fractions*. Note that the fractions are not pure and contain mixtures of hydrocarbons within a range of boiling points. The fractionating tower, shown below, shows the temperature gradient which separates the fractions according to their boiling points.

Fractions with the lowest boiling points are collected at the top of the tower.

On descending the tower, the temperature rises and the boiling points increase.

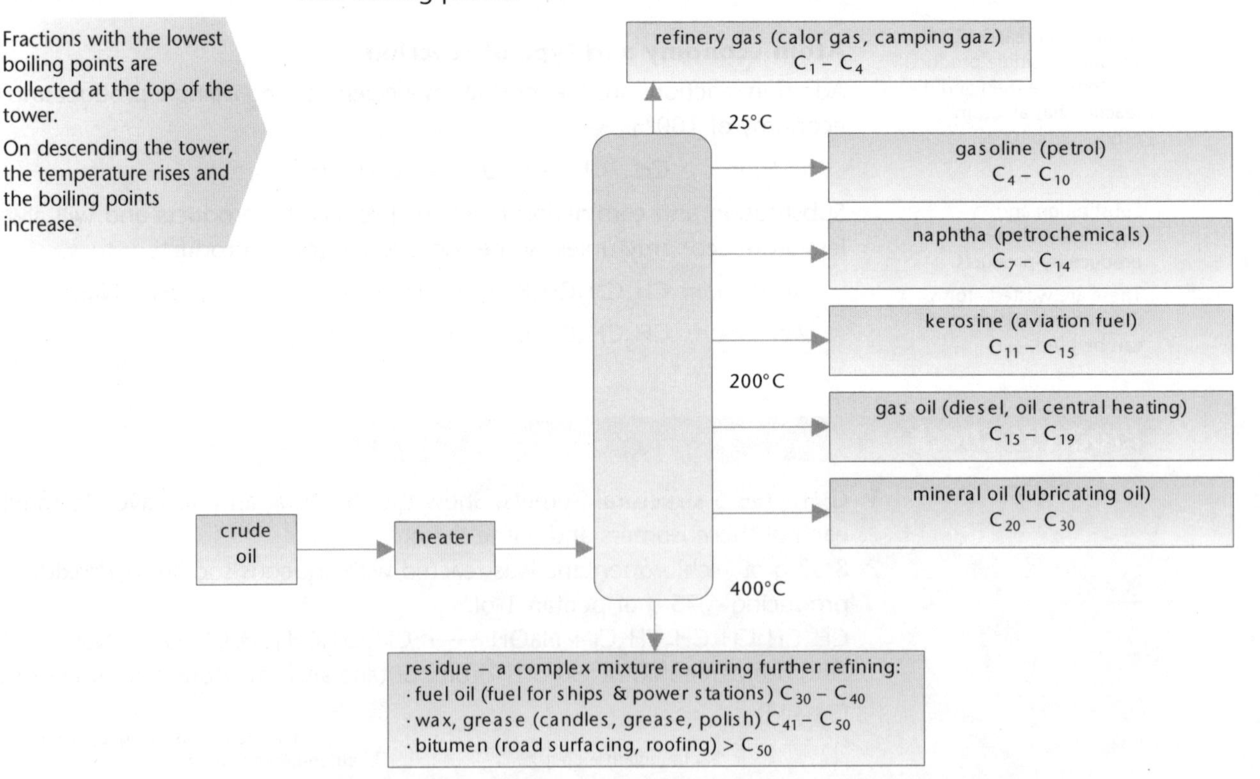

Cracking

Cracking is the starting point for the manufacture of many organics.

Cracking is the breaking down of an unsaturated hydrocarbon into smaller hydrocarbons. Cracking breaks C–C bonds in alkanes.

The purpose of cracking is to produce high demand hydrocarbons:

- short-chain alkanes for use in petrol
- alkenes, as a feedstock for a wide range of organic chemicals, including polymers (see also page 109).

Thermal cracking heats the *naphtha fraction* with steam at a high temperature (about 800°C) and high pressure. This forms a mixture of straight-chain alkanes and alkenes (mainly ethene) with a small proportion of branched and cyclic hydrocarbons. Some hydrogen is also produced.

The following equations summarise the overall process of cracking.

$$C_{10}H_{22} \longrightarrow C_8H_{18} + C_2H_4$$
$$C_{10}H_{22} \longrightarrow C_6H_{14} + 2C_2H_4$$
$$\text{alkane} \longrightarrow \text{alkane} + \text{alkene}$$

There are many more equations possible but all must produce both an alkane and an alkene.

Thermal cracking breaks C–C bonds by homolytic fission forming radicals.

Catalytic cracking breaks C–C bonds by heterolytic fission forming ions.

Catalytic cracking processes heavy long-chain fractions obtained in larger quantities than required. A mixture of the vapourised fraction and a zeolite catalyst are reacted at about 450°C using a slight pressure only. The process forms a higher proportion of branched and cyclic hydrocarbons than thermal cracking (see also reforming and isomerisation below). These are used for petrol.

Reforming

In a car engine, petrol has the tendency to auto-ignite before the spark, causing 'knocking'. Resistance to auto-ignition is measured as the **octane number** of the fuel.

Straight-chain hydrocarbons have lower octane numbers than branched-chain and cyclic hydrocarbons.

Reforming uses heat and pressure to convert unbranched fractions into cycloalkanes (e.g. cyclohexane) and arenes (e.g. benzene). These products are used in petrol and as a feedstock for a wide range of organic chemicals including many pharmaceuticals and dyes.

Examples

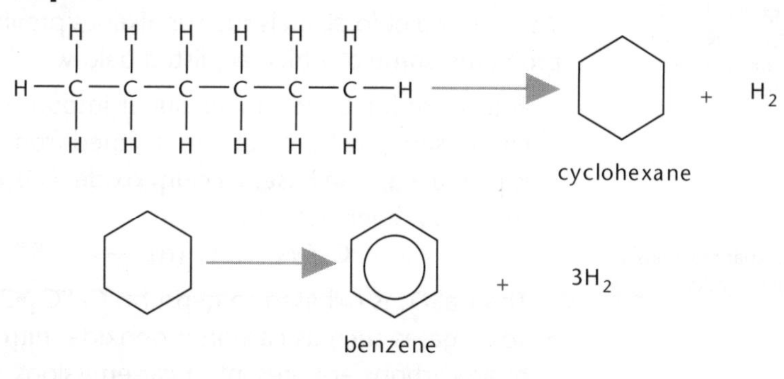

Unleaded petrol and lead-replacement petrol require a higher proportion of branched and cyclic hydrocarbons for effective combustion. This has increased demand for reformed alkanes.

Isomerisation

Chemists are improving fuels by use of oxygenates such as methanol and ethanol. Hydrogen is also being developed as a fuel to replace petrol in the future.

Isomerisation converts unbranched hydrocarbons into branched hydrocarbons, needed for petrol.

Example

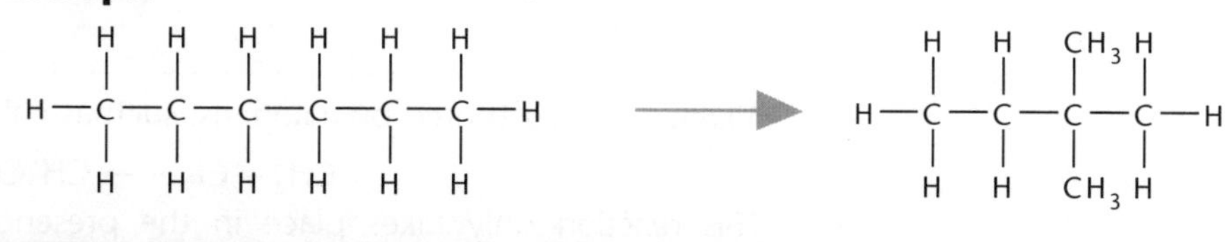

Progress check

1 Write down the formula of an alkane present in each of the main fractions obtained from crude oil.

2 Construct equations for two possible reactions taking place during the cracking of $C_{14}H_{30}$.

1 Any alkane with a formula C_nH_{2n+2} (see page 125) to match each fraction.

2 Any 2 equations alkane ⟶ alkane + alkene, e.g. $C_{14}H_{30} \longrightarrow C_8H_{18} + C_6H_{12}$

4.3 Alkanes

After studying this section you should be able to:

- *understand that combustion of alkanes provides useful energy*
- *describe the substitution reactions of alkanes with halogens*

LEARNING SUMMARY

Alkanes

General formula: C_nH_{2n+2}

AQA M1

Alkanes are generally unreactive. Alkanes contain only C–H and C–C bonds, which are relatively strong and difficult to break. The similar electronegativities of carbon and hydrogen give molecules which are non-polar. Alkanes are the typical 'oils' used in many non-polar solvents and they do not mix with water.

Combustion of alkanes

AQA M1

Combustion with oxygen is the most important reaction of alkanes giving their immediate use as fuels:

natural gas: $CH_4 + 2O_2 \longrightarrow CO_2 + 2H_2O$

petrol: $C_8H_{18} + 12\frac{1}{2}O_2 \longrightarrow 8CO_2 + 9H_2O$

Although generally unreactive, alkanes do react with oxygen and the halogens.

The burning of fossil fuels such as alkanes provides society with many environmental problems, some of which are listed below.

- Petroleum fractions contain sulfur impurities. Unless removed before combustion, sulfur oxides are formed from which acid rain (H_2SO_3 and H_2SO_4) is produced. The base, calcium oxide, is used to remove sulfur oxides from flue gases in power stations:

  $$CaO(s) + SO_2(g) \longrightarrow CaSO_3(s)$$

  The $CaSO_3$ is oxidised to gypsum, $CaSO_4{\bullet}2H_2O$, which is used to make plaster.
- Toxic gases such as carbon monoxide, nitrogen oxides and unburnt hydrocarbons are present in car emissions. The development of catalytic converters has helped to remove these polluting gases.
- Excessive combustion of fossil fuels increases carbon dioxide emissions contributing to global warming.

Remember that reactions with oxygen produce oxides.

Carbon dioxide, methane and water vapour are the main greenhouse gases that contribute to global warming (see page 121).

Substitution of alkanes with halogens

AQA M2

> A *substitution reaction* involves the swapping over of one species for another.
>
> KEY POINT

Alkanes are substituted by halogens, such as chlorine and bromine.

$$CH_4 + Cl_2 \longrightarrow CH_3Cl + HCl$$

This reaction only takes place in the presence of ultraviolet radiation, which produces highly reactive free radicals.

> A free radical, such as $Cl\bullet$, $CH_3\bullet$,
> - is a highly reactive species with an unpaired electron
> - reacts by pairing of an unpaired electron
> - is often involved in chain reactions.
>
> KEY POINT

Mechanism of free radical substitution

Initiation

Make sure that you learn this mechanism – these are easy marks in exams if you do!

In this stage, the reaction starts.

Ultraviolet radiation provides energy to break Cl–Cl bonds homolytically producing chlorine free radicals, Cl•

$$Cl_2 \longrightarrow 2Cl\bullet$$

Propagation

In this stage, the reaction products are made.

The chlorine free radicals catalyse the reaction.

See also the breakdown of ozone, pages 122–123.

Free radicals are recycled in a chain reaction

$$CH_4 + Cl\bullet \longrightarrow CH_3\bullet + HCl$$
$$CH_3\bullet + Cl_2 \longrightarrow CH_3Cl + Cl\bullet$$

Termination

In this stage, free radicals react together and are removed from the reaction mixture.

$$Cl\bullet + Cl\bullet \longrightarrow Cl_2$$
$$CH_3\bullet + Cl\bullet \longrightarrow CH_3Cl$$
$$CH_3\bullet + CH_3\bullet \longrightarrow CH_3CH_3$$

Impure products

This is answered poorly in exams.

Further substitution of the reaction products is possible, producing a mixture of products:

$$CH_3Cl \xrightarrow{Cl\bullet} CH_2Cl_2 \xrightarrow{Cl\bullet} CHCl_3 \xrightarrow{Cl\bullet} CCl_4$$

Progress check

1 Write an equation for the complete combustion of butane.

2 Butane reacts with chlorine in a substitution reaction.
   (a) What conditions are essential for this reaction? Explain your answer.
   (b) Write an equation for this reaction.
   (c) Write the structural formula of 2 possible monosubstituted products.
   (d) What is the formula for the final product obtained from complete substitution?

1 $C_4H_{10}(g) + 6\frac{1}{2}O_2(g) \longrightarrow 4CO_2(g) + 5H_2O(l)$

2 (a) UV; to generate radicals from Cl_2
   (b) $C_4H_{10} + Cl_2 \longrightarrow C_4H_9Cl + HCl$
   (c) Any 2 isomers of C_4H_9Cl, e.g. $CH_3CH_2CH_2CH_2Cl$ and $CH_3CH_2CHClCH_3$
   (d) C_4Cl_{10}

4.4 Alkenes

After studying this section you should be able to:

- *recognise the nature of E/Z isomerism in alkenes*
- *understand that the double bond in alkenes takes part in addition reactions with electrophiles*
- *understand the use of alkenes in the production of organic compounds*
- *understand what is meant by addition polymerisation*

LEARNING SUMMARY

Alkenes

General formula: C_nH_{2n}

AQA M2

Alkenes are unsaturated compounds with a C=C double bond. The high electron density of the double bond makes alkenes more reactive than alkanes.

Naming of alkenes

Position isomerism
A type of structural isomerism in which the functional group is in a different position on the carbon skeleton.

The position of the double bond is indicated using one number only, as shown in the example below for two isomers of C_4H_8.

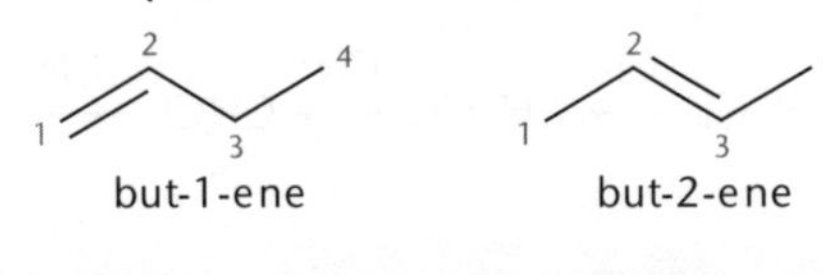

Only one number is needed. In but-1-ene, a double bond starting at carbon-1 must finish at carbon-2

What is a double bond?

The coverage here is greatly simplified but sufficient for AS Chemistry.

In saturated compounds, the C–H and C–C single bonds are σ-bonds. A σ-bond is on the C–H or C–C axis, formed by overlap of s- and p- atomic orbitals.

In unsaturated compounds, the C=C double bond comprises a σ- and a π-bond.

A π-bond is above and below the C–C axis, formed by overlap of atomic p-orbitals. The π-bond introduces an area of high electron density in the molecule, shown in the diagram of ethene below.

σ- and π-bonds are *molecular orbitals* – these bond together the atoms in a molecule.

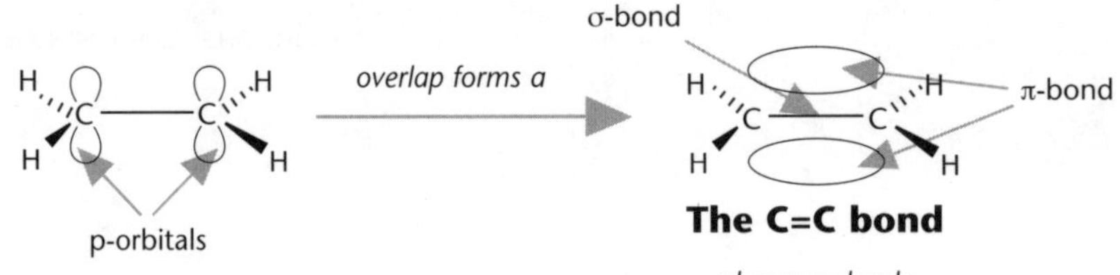

planar molecule

3 σ-electron pairs around each carbon atom.
Bond-angle is approximately 120°

E/Z isomerism

AQA M2

Stereoisomers are molecules with the same structural formula but with a different arrangement in space.

KEY POINT

See also structural isomerism, page 98.

E/Z isomerism is a type of stereoisomerism that is present in some unsaturated hydrocarbons with a C=C bond.

Single C–C bonds can rotate. The C=C double bond, however, restricts rotation and prevents groups from moving from one side of the double bond to the other. This can give rise to *E/Z* isomerism.

For *E/Z* isomerism, there must be:

No rotation possible about a C=C bond.

- a C=C double bond
- two **different** groups attached to **each** carbon end of the double bond.

These are relatively easy exam marks. Make sure you learn this thoroughly.

If two of the groups are the same, this is also referred to as *cis-trans* isomerism:

- the *E* isomer (*trans* isomer) has the same groups on opposite sides
- the *Z* isomer (*cis* isomer) has the same groups on the same side.

You should be able to label *E* and *Z* isomers in compounds that have *cis* and *trans* isomers.

The *E* and *Z* isomers of but-2-ene are shown below:

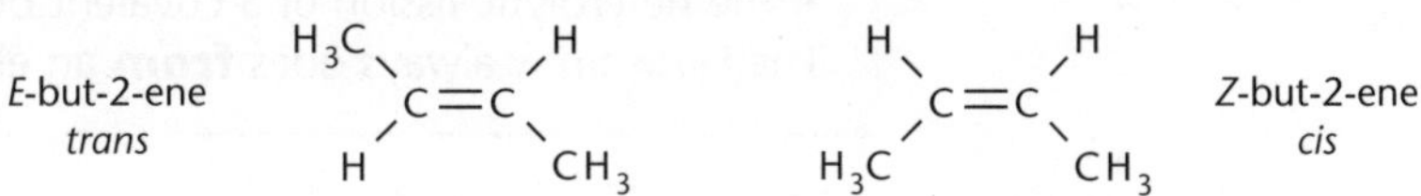

Cis-trans isomerism is limited to unsaturated compounds in which two of the different groups are the same (as in the example above). The *E/Z* naming system can be applied to compounds with more than 2 different groups attached to either end of the C=C double bond.

In exams, you would be expected to show *E/Z* isomers for this example but you would not be expected to label which is which.

The *E/Z* isomers of 3-methylhept-3-ene are shown below:

H, CH_3 / C=C / $CH_3CH_2CH_2$, CH_2CH_3 H, CH_2CH_3 / C=C / $CH_3CH_2CH_2$, CH_3

It is impossible to label these as *cis* and *trans* isomers. The rules for labelling each stereoisomer in this example as *E* and *Z* go beyond the demands of AS Level Chemistry.

Addition reactions of alkenes

AQA M2

An addition reaction of an alkene involves the opening of the double bond with the formation of a saturated addition product.

> **KEY POINT**
> An *addition reaction* involves two species adding together to make one. Addition usually converts an **unsaturated** compound into a **saturated** compound.

This is made possible by the high electron density of the π-bond which attracts **electrophiles**.

> **KEY POINT**
> An *electrophile*, such as Br_2, HBr, H_2SO_4, NO_2^+:
> - is an electron-deficient species
> - 'attacks' an electron-rich carbon atom or double bond by **accepting a pair of electrons**.

This type of reaction is called **electrophilic addition**.

Addition of bromine

The addition reaction between an alkene and bromine, in which the orange colour of bromine is decolourised, is used as a test for unsaturation.

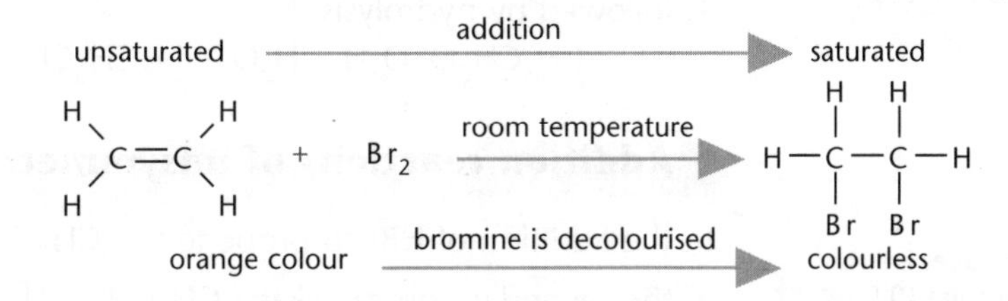

Mechanism of electrophilic addition

Be careful with the direction of the curly arrows.

Don't get confused by the charges and partial charges within this mechanism.

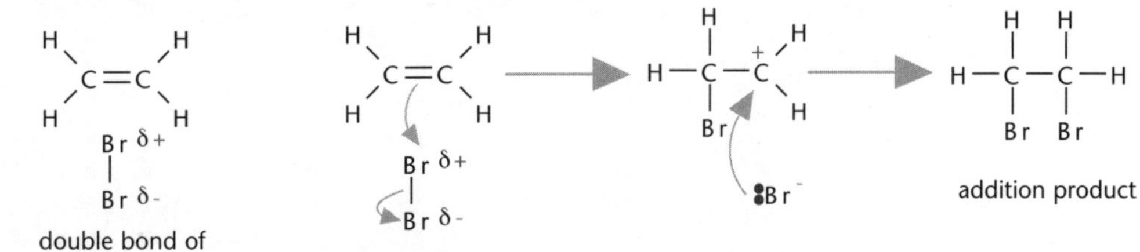

Remember that a curly arrow shows the movement of an electron pair.

> In mechanisms, *a curly arrow* is used to show the movement of **a pair of electrons**.
> The movement of the electron pair involves either:
> - the formation of a new covalent bond or
> - the heterolytic fission of a covalent bond (see page 100).
>
> The *curly arrow* always goes **from** an electron pair (or double bond).
>
> KEY POINT

Reduction with hydrogen

> Reduction of an organic compound involves loss of oxygen OR gain of hydrogen (accompanied by a gain of electrons).
>
> KEY POINT

Hydrogenation of C=C is used in the production of solid margarine from unsaturated liquid vegetable oils. Saturated fats are solid and the amount of hydrogenation can be controlled to give the correct texture of margarine.

Alkenes are reduced when treated with hydrogen gas in the presence of a nickel catalyst at 150°C. The process of adding hydrogen across a double bond is sometimes referred to as **hydrogenation**.

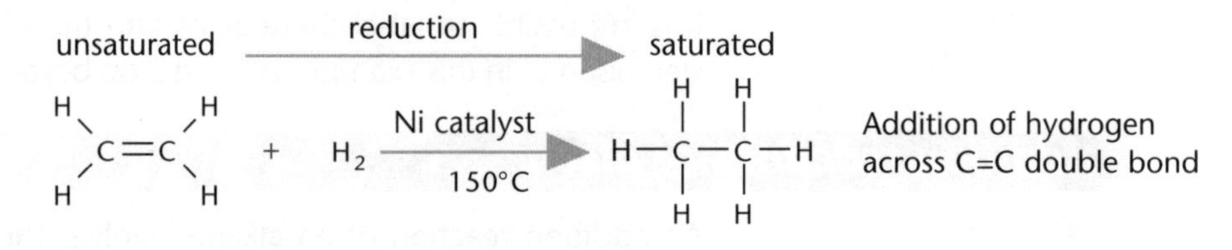

Further addition reactions of ethene

Addition reactions of alkenes are useful in organic synthesis. Some further addition reactions of ethene are shown in the flowchart below.

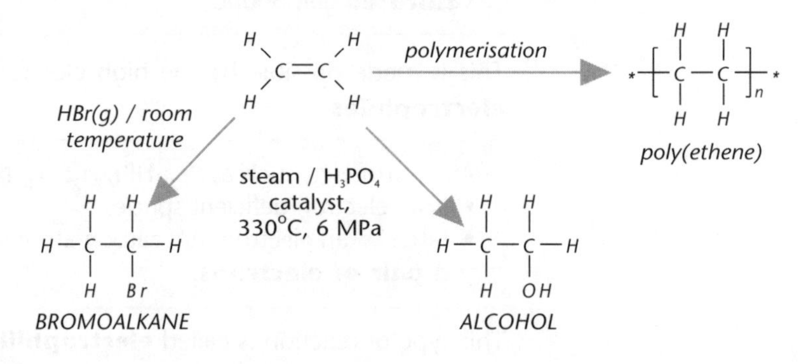

For more details: polymerisation, see below; preparation of alcohols, see also page 112.

Although ethanol is made industrially using steam and phosphoric acid H_3PO_4, an alternative synthesis uses:

- addition with concentrated sulfuric acid at room temperature
  $$C_2H_4 + H_2SO_4 \longrightarrow CH_3CH_2OSO_3H$$
- followed by hydrolysis
  $$CH_3CH_2OSO_3H + H_2O \longrightarrow CH_3CH_2OH + H_2SO_4$$

Addition reactions of unsymmetrical alkenes

These structures are shown on page 109.

In the addition of HBr to propene $CH_3CH{=}CH_2$, two products are possible:

- the secondary bromoalkane $CH_3CHBrCH_3$ as the major product
- the primary bromoalkane $CH_3CH_2CH_2Br$ as the minor product.

Mechanism

The driving force for the formation of $CH_3CHBrCH_3$ is the intermediate carbocation, shown in the mechanism below.

You need to know this mechanism, but you also need to know how to explain it in terms of carbocation stabilities

The reaction with H_2SO_4 proceeds via a similar mechanism. H_2SO_4 can be treated as H-OSO_2OH.

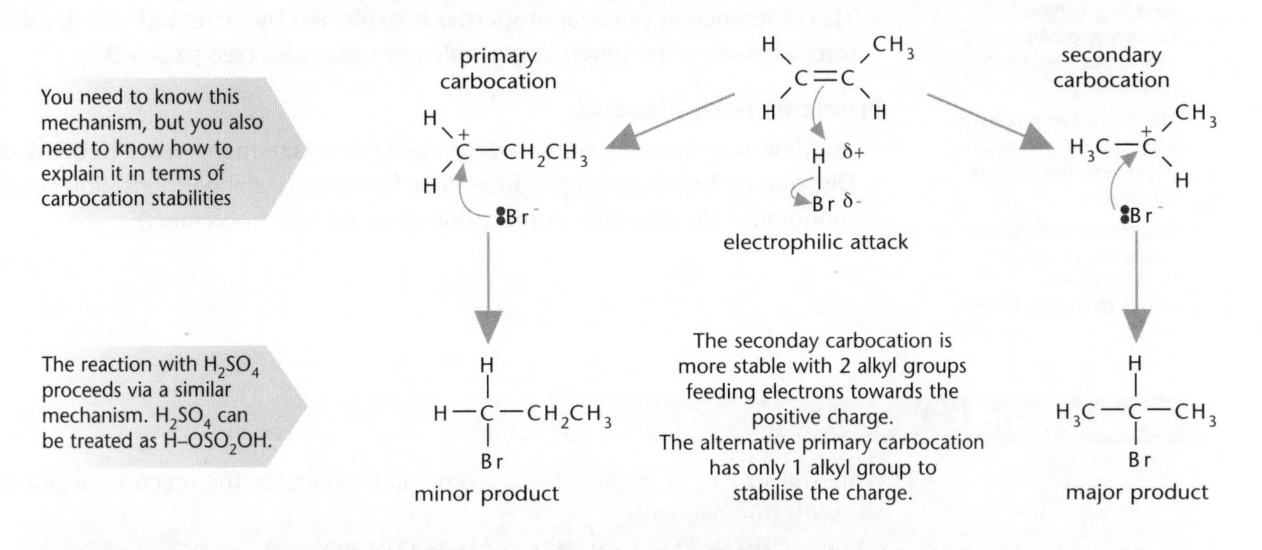

Addition polymerisation of alkenes

AQA M2

An addition polymer is a long-chain molecule with a high molecular mass, made by joining together many small molecules called monomers:

MANY MONOMERS ⟶ SINGLE POLYMER

n molecules ⟶ 1 single molecule

- The **monomer** used is an **unsaturated** alkene containing a double C=C bond.
- The addition **polymer** formed is a **saturated** compound **without** a double bond.

Many different addition polymers can be formed by using different alkene monomer units, based on the ethene molecule.

The equations below show the formation of poly(ethene), poly(chloroethene) (*pvc*) and poly(tetrafluoroethene) (*PTFE*).

You should be able to draw a short section of a polymer given the monomer units (and *vice versa*).

Make sure that you can show the repeat unit of a polymer.

- Notice that the principle is the same for each addition polymerisation.
- Each of the polymer structures shows the *repeat unit* in brackets. This is repeated thousands of times in each polymer molecule.
- It is important to show the repeat unit with 'side-links' to indicate that both sides are also attached to other repeat units.

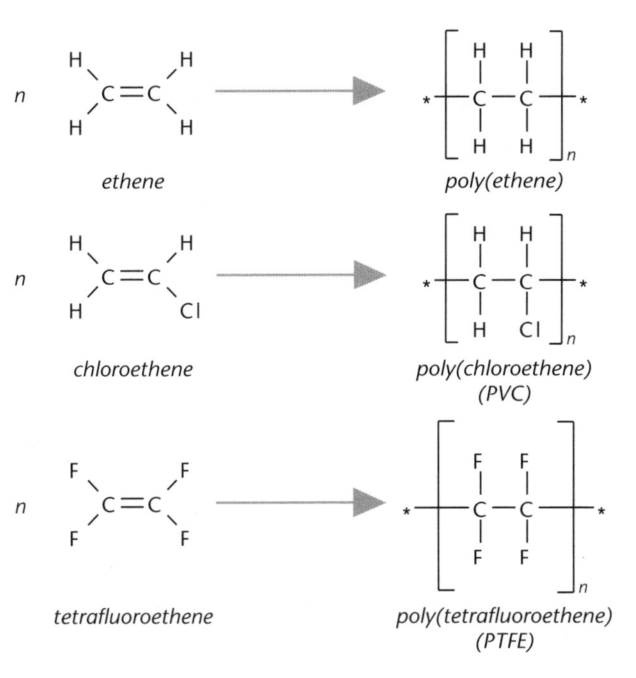

Rather than simply disposing of waste polymers, there is now movement towards recycling, combustion for energy production, and use as a feedstock for cracking.

This eliminates problems with disposal and helps to preserve finite energy resources.

Degradable polymers are also being developed from renewable resources such as corn starch.

Properties of monomers and polymers

- The monomers are volatile liquids or gases.
- Polymers are solids.
- This difference in physical properties is explained by increased van der Waals' forces between the much larger polymer molecules (see page 49).

Problems with disposal

- Addition polymers are non-biodegradable and take many years to break down.
- Disposal by burning can produce toxic fumes (e.g. depolymerisation produces monomers; dioxins from combustion of chlorinated polymers).

Progress check

1 Write the structural formula for the organic product for the reaction of but-2-ene with the following:
(a) Br_2; (b) H_2/Ni; (c) HBr; (d) H_2O/H_3PO_4.

2 Show the repeat unit for the addition polymer formed from propene.

1 (a) $CH_3CHBrCHBrCH_3$ (b) $CH_3CH_2CH_2CH_3$ (c) $CH_3CH_2CHBrCH_3$
(d) $CH_3CH_2CHOHCH_3$.

2

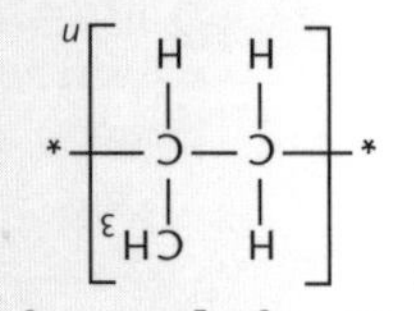

4.5 Alcohols

After studying this section you should be able to:

LEARNING SUMMARY

- *understand the polarity and physical properties of alcohols*
- *describe how ethanol is prepared*
- *describe oxidation reactions of alcohols*
- *describe esterification and dehydration of alcohols*
- *describe tests for the presence of the hydroxyl group*

Alcohols

General formula: $C_nH_{2n+1}OH$

AQA M2

Types and naming of alcohols

The functional group in alcohols is the **hydroxyl** group, C–OH.

Alcohols can be classified as primary, secondary or tertiary, depending on how many alkyl groups are bonded to C–OH. You can see how these types of alcohols are named in the diagrams below.

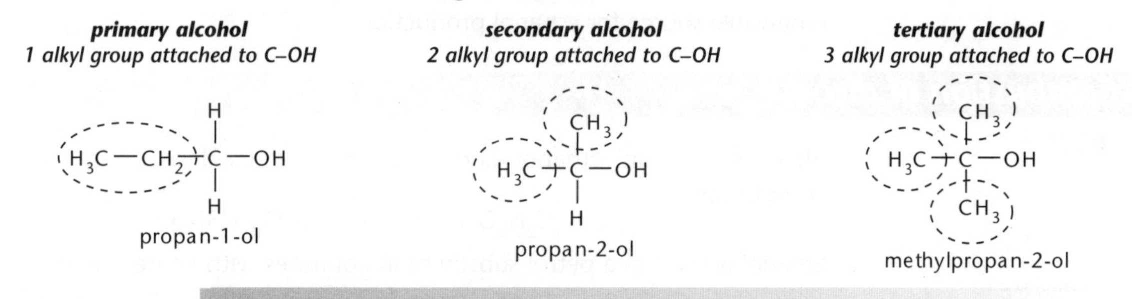

Polarity of alcohols

Carbon, oxygen and hydrogen have different electronegativities, and alcohols have polar molecules:

Oxygen is more electronegative than both carbon and hydrogen – both carbon and hydrogen are electron deficient.

The polarity produces electron-deficient carbon and hydrogen atoms, indicated in the diagram above. Depending upon the reagents used, alcohols can react by breaking either the C–O or O–H bond.

The properties of alcohols are dominated by the hydroxyl group, C–OH.

hydrogen bonding between ethanol molecules

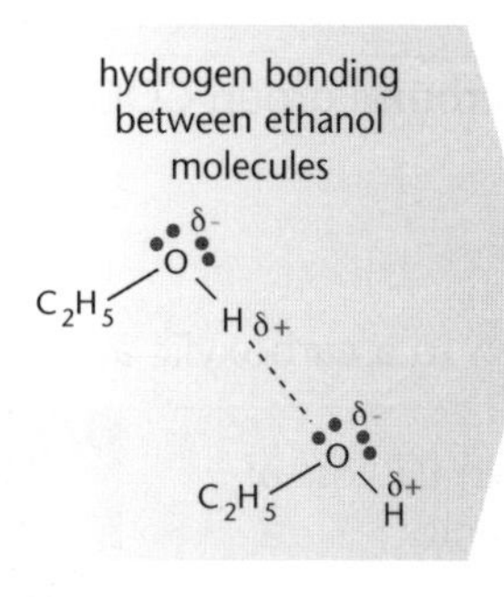

Physical properties of alcohols

The –OH group dominates the physical properties of short-chain alcohols.

Hydrogen bonding takes place between alcohol molecules, resulting in:

- higher melting and boiling points than alkanes of comparable M_r
- solubility in water.

The solubility of alcohols in water decreases with increasing carbon chain length as the non-polar contribution to the molecule becomes more important.

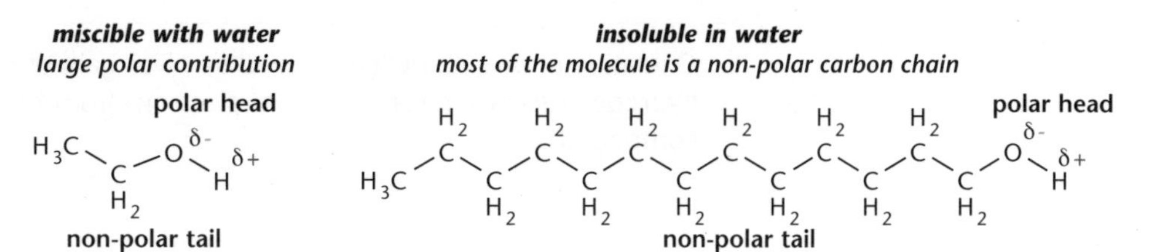

Preparation of ethanol

AQA M2

Ethanol finds widespread uses: in alcoholic drinks, as a solvent in the form of methylated spirits, and as a fuel. There are two main methods for its production.

Fermentation of sugars (for alcoholic drinks)

Yeast is added to an aqueous solution containing sugars.

E.g. $$C_6H_{12}O_6(aq) \longrightarrow 2C_2H_5OH(aq) + 2CO_2(g)$$

Hydration of ethene (for industrial alcohol)

Ethene and steam are passed over a phosphoric acid catalyst at 330°C under high pressure (see page 108).

$$C_2H_4(g) + H_2O(g) \longrightarrow C_2H_5OH(l)$$

The preparation method used for ethanol may depend on the raw materials available.

An oil-rich country may use oil to produce ethene, then ethanol. However, this method uses up finite oil reserves.

In a country without oil reserves (especially in hot climates), sugar may provide a renewable source for ethanol production.

Combustion of alcohols

AQA M2

Alcohols such as ethanol and methanol are used as fuels, making use of combustion.

$$C_2H_5OH + 3O_2 \longrightarrow 2CO_2 + 3H_2O$$

Methanol is an 'oxygenate'. The oxygen in its formula aids combustion.

Ethanol is used as a petrol substitute in countries with limited oil reserves.

Methanol is used as a petrol additive in the UK to improve combustion of petrol. It also has increasing importance as a feedstock in the production of organic chemicals.

Oxidation of alcohols

AQA M2

The oxidation of different alcohols is an important reaction in organic chemistry. It links alcohols with aldehydes, ketones and carboxylic acids, shown below.

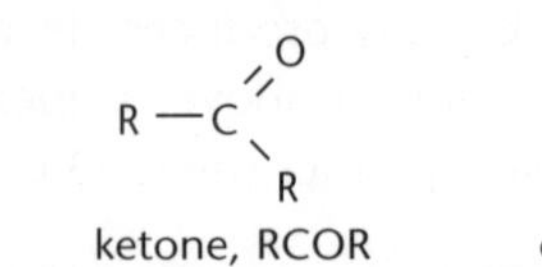

aldehyde, RCHO ketone, RCOR carboxylic acid, RCOOH

- Aldehydes and ketones both contain the **carbonyl group** C=O.
- Carboxylic acids contain the **carboxyl** group COOH.

The stages of oxidation are shown below:

hydroxyl group C–OH ⟶ carbonyl group C=O ⟶ carboxyl group COOH

Oxidation of primary alcohols

A primary alcohol can be oxidised to an aldehyde and then to a carboxylic acid.

> **KEY POINT**
> Oxidation of an organic compound involves gain of oxygen OR loss of hydrogen (accompanied by loss of electrons from the organic compound).

The stages of oxidation are shown below:

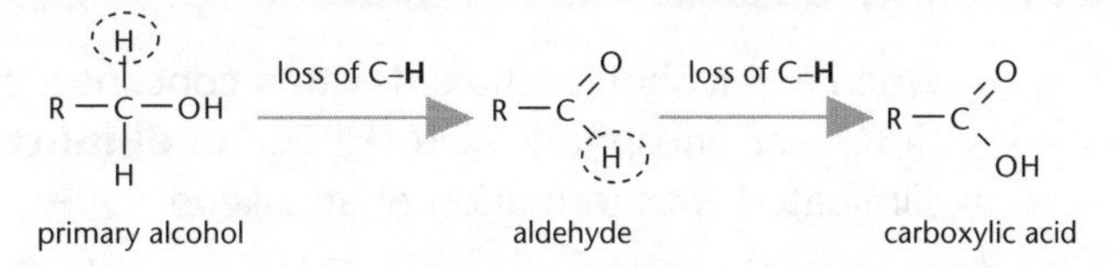

This is carried out using an oxidising agent. The oxidising agent used is a mixture of concentrated sulfuric acid, H_2SO_4 (source of H^+) and potassium dichromate, $K_2Cr_2O_7$ (source of $Cr_2O_7^{2-}$).

For balanced equations, the oxidising agent can be shown simply as [O].

- By heating and distilling the product immediately, oxidation can be stopped at the aldehyde stage.

$$CH_3CH_2OH + [O] \longrightarrow CH_3CHO + H_2O$$

- By refluxing with an excess of the oxidising agent, further oxidation takes place to form the carboxylic acid.

$$CH_3CH_2OH + 2[O] \longrightarrow CH_3COOH + H_2O$$

The orange dichromate ions, $Cr_2O_7^{2-}$, are reduced to green Cr^{3+} ions.

Oxidation of secondary alcohols

By heating with $H^+/Cr_2O_7^{2-}$ a secondary alcohol can be oxidised to form a ketone. No further oxidation normally takes place.

secondary alcohol $\longrightarrow$ ketone

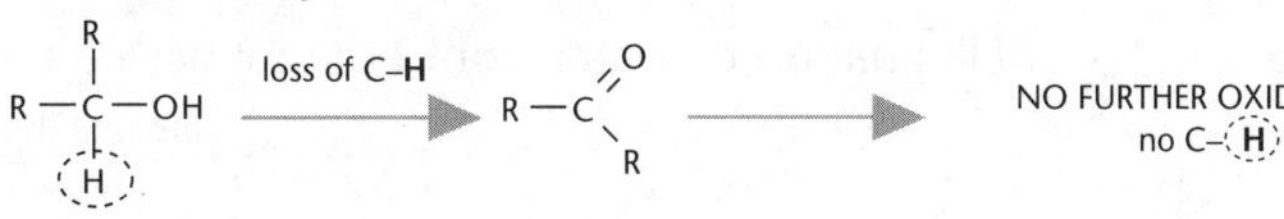

The equation for the oxidation of propan-2-ol is shown below.

$$CH_3(CHOH)CH_3 + [O] \longrightarrow CH_3COCH_3 + H_2O$$

Oxidation of tertiary alcohols

Tertiary alcohols are not oxidised under normal conditions.

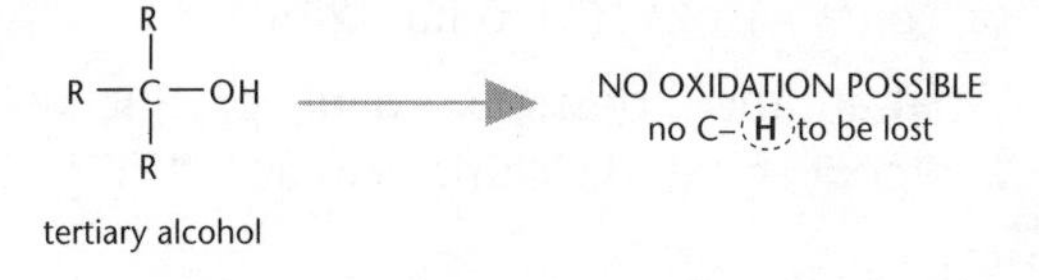

Distinguishing between aldehydes and ketones

Tollens' reagent is a solution of silver nitrate (as a source of Ag^+ ions) in ammonia.
$Ag^+ \longrightarrow Ag$

Fehling's solution is an alkaline solution of Cu^{2+} ions.
$Cu^{2+} \longrightarrow Cu^+$

The presence of an aldehyde can easily be detected using two tests.

Heat aldehyde with Tollens' reagent

- Tollens' reagent $\longrightarrow$ silver mirror

Heat aldehyde with Fehling's solution

- Benedict's / Fehling's solution $\longrightarrow$ brick-red precipitate of Cu_2O

Ketones cannot be oxidised and **do not** react with the above solutions and this provides an easy method for distinguishing between an aldehyde and a ketone.

Dehydration of alcohols

AQA M2

When an alcohol is refluxed with a concentrated acid catalyst such as sulfuric acid H_2SO_4, or phosphoric acid H_3PO_4, an **elimination** reaction takes place: water is eliminated with formation of an alkene.

$$C_2H_5OH \longrightarrow C_2H_4 + H_2O$$

$$H-\overset{H}{\underset{H}{C}}-\overset{H}{\underset{H}{C}}-OH \xrightarrow[\text{reflux}]{\text{conc. } H_2SO_4} \begin{matrix} H & & H \\ & C{=}C & \\ H & & H \end{matrix} + H_2O$$

Use hot concentrated acid for elimination of water from an alcohol.

Elimination (1 ⟶ 2) is the opposite process to addition (2 ⟶ 1).

KEY POINT

An *elimination* reaction involves one species breaking up into two – the opposite to addition.

Elimination usually converts a **saturated** compound into an **unsaturated** compound.

Alkenes, such as ethene, can be synthesised from renewable resources rather than from crude oil. Sugars, such as glucose, can be fermented to produce ethanol, C_2H_5OH.

See page 112 for more details of methods used to prepare ethanol.

E.g.

$$\underset{\text{glucose}}{C_6H_{12}O_6} \longrightarrow 2C_2H_5OH + 2CO_2$$

Elimination of water from ethanol with an acid catalyst produces ethene.

$$C_2H_5OH \xrightarrow[\text{heat}]{H^+ \text{ catalyst}} C_2H_4 + H_2O$$

The alkenes can then be polymerised to form plastics.

Progress check

1 Why does ethanol dissolve in water?

2 Write equations to show possible oxidations for the following:
(a) butan-1-ol; (b) butan-2-ol.

3 What are the 2 possible organic products of the reaction between butan-2-ol and concentrated sulfuric acid at 170°C?

1 Alcohol –OH group forms H-bonds with water molecules.

2 (a) $CH_3CH_2CH_2CH_2OH + [O] \longrightarrow CH_3CH_2CH_2CHO + H_2O$
$CH_3CH_2CH_2CH_2OH + 2[O] \longrightarrow CH_3CH_2CH_2COOH + H_2O$
(b) $CH_3CH_2CHOHCH_3 + [O] \longrightarrow CH_3CH_2COCH_3 + H_2O$

3 $CH_3CH{=}CHCH_3$ and $CH_3CH_2CH{=}CH_2$

4.6 Haloalkanes

After studying this section you should be able to:

- understand the nature of the polarity in haloalkanes
- describe nucleophilic substitution reactions of haloalkanes
- explain the relative rates of hydrolysis of different haloalkanes
- describe elimination reactions of haloalkanes
- understand that reaction conditions can be used to control the products

LEARNING SUMMARY

Haloalkanes

General formula: $C_nH_{2n+1}X$, where X = F, Cl, Br or I

AQA M2

Types and naming of haloalkanes

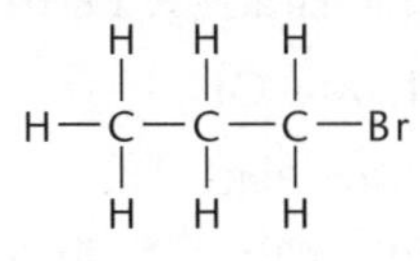

1-bromopropane *primary*

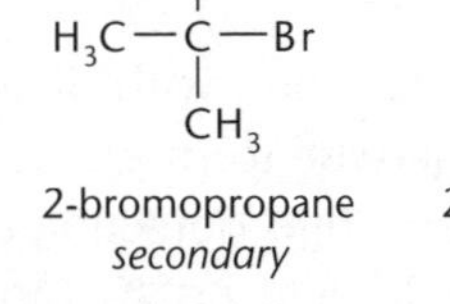

2-bromopropane *secondary*

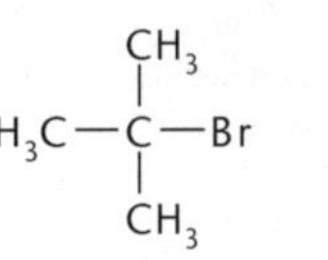

2-bromo-2-methylpropane *tertiary*

Uses of haloalkanes

Chlorofluorocarbons, CFCs were used as refrigerants, solvents, propellants, dry cleaning and degreasing agents, and for blowing polystyrene.

CFCs are now known to deplete the ozone layer. Their use is now banned in countries that have signed up to the Montreal Protocol.

Chloroethene and tetrafluoroethene are used to produce the plastics PVC and PTFE.

Haloalkanes are used as synthetic intermediates in chemistry and solvents.

Polychlorinated compounds are used as herbicides.

Polarity of haloalkanes

Carbon and halogens have different electronegativities and halogenoalkanes have polar molecules with a polar C–X bond.

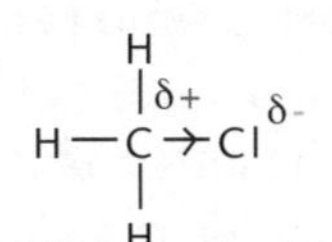

chlorine is more electronegative than carbon
electron flow from carbon to chlorine – dipole produced

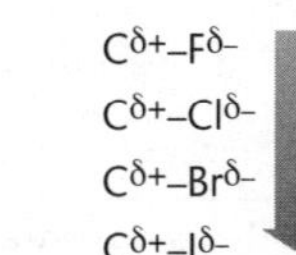

polarity **decreases**

The polarity produces an electron-deficient carbon atom, $C^{\delta+}$ which is important in the reactions of haloalkanes. The polarity decreases from fluorine to iodine, reflecting the decrease in electronegativity down the halogen group.

Nucleophilic substitution reactions of halogenoalkanes

AQA M2

The electron-deficient carbon atom of the polar $C^{\delta+}$–$X^{\delta-}$ bond attracts **nucleophiles**, allowing a **nucleophilic substitution** reaction to take place in which the nucleophile replaces the halogen atom.

The hydrolysis of halogenoalkanes

A **hydrolysis** reaction is a reaction in which water breaks a bond.

KEY POINT

Reaction with aqueous hydroxide ions, $OH^-(aq)$ produces an alcohol.

$$C_2H_5Br + OH^-(aq) \xrightarrow[\textit{reflux}]{OH^-/H_2O} C_2H_5OH + Br^-$$

The hydroxide ion, OH^-, behaves as a nucleophile.

Mechanism of nucleophilic substitution with primary halogenoalkanes

Primary halogenoalkanes react in a **one-step** mechanism.

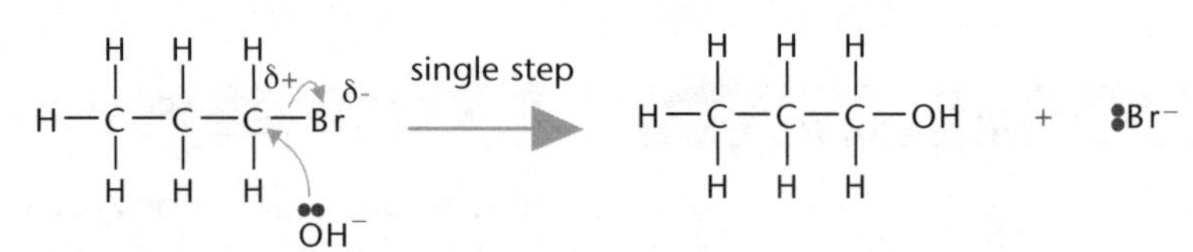

NaOH(aq) is used as a source of OH^- (aq).

The OH^- ion behaves as a nucleophile by donating an electron pair.

Note the usefulness of the curly arrow in showing movement of an electron pair:

- the $:OH^-$ nucleophile donates its electron pair to the electron-deficient $C^{\delta+}$ of the bromoalkane, forming a covalent bond;
- the C–Br bond is broken by heterolytic fission with loss of its electron pair to form a $:Br^-$ ion.

Comparing the hydrolysis reactions of haloalkanes

An alternative method uses water itself for the hydrolysis. The haloalkane is heated with a mixture of water, ethanol and $AgNO_3$(aq). The rate of hydrolysis is monitored directly by observing precipitation of replaced halide ions as silver halides.

The rates of hydrolysis of different haloalkanes can be investigated as follows:

- Alkaline hydrolysis with NaOH(aq)/H_2O, reflux:

$$R\text{–}X + OH^- \longrightarrow R\text{–}OH + X^-$$

- Acidification with dilute nitric acid. This removes excess NaOH, which would otherwise form a precipitate with the silver nitrate in the next stage, preventing detection of any silver halides.
- Addition of $AgNO_3$(aq) shows the presence of aqueous halide ions (see also Testing for halide ions, page 64).

$$Ag^+(aq) + X^-(aq) \longrightarrow AgX(s)$$

The intensity of any precipitate indicates the extent and rate of the hydrolysis.

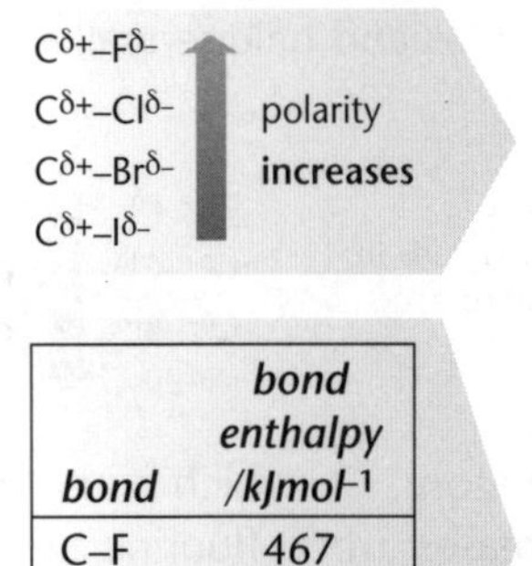

bond	*bond enthalpy /kJmol⁻¹*
C–F	467
C–Cl	364
C–Br	290
C–I	228

KEY POINT

Two factors – which is more important?

- **Polarity** predicts that the more polar C–F bond would attract water most easily and would give the fastest reaction.
- **Bond enthalpy** predicts that the C–I bond would be broken most easily and would give the fastest reaction.

For this reaction, bond enthalpy is more important than polarity.
The rate of hydrolysis increases as the C–X bond weakens.

Further examples of nucleophilic substitution

AQA M2

Nucleophilic substitution of halogenoalkanes is used in organic synthesis to introduce different functional groups onto the carbon skeleton. The reaction scheme below shows how haloalkanes react with ammonia and cyanide ions.

For hydrolysis, water is used as the solvent. For other nucleophilic substitutions, it is important that water is absent or hydrolysis will take place.

Ethanol is used as the alternative solvent.

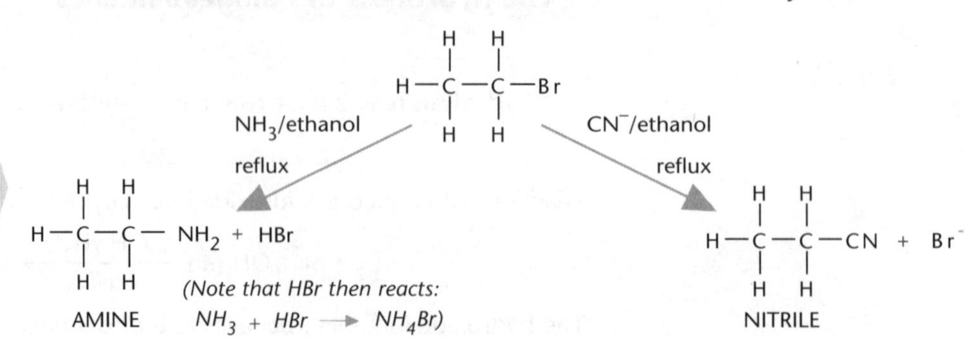

The ability of haloalkanes to undergo nucleophilic substitution is important in organic synthesis as compounds with different functional groups can be synthesised.

Elimination reactions of haloalkanes

AQA M2

When a haloalkane is refluxed with hydroxide ions in **anhydrous** conditions (using NaOH in **ethanol** as a solvent at 78°C), an **elimination** reaction takes place.

$$H_3C{-}CH_2Br + OH^- \xrightarrow[\text{reflux}]{OH^-/\text{ethanol}} H_2C{=}CH_2 + Br^- + H_2O$$

The hydroxide OH^- ion behaves as a base, accepting a proton H^+ to form H_2O.

Elimination versus hydrolysis

AQA M2

The reaction of hydroxide ions with haloalkanes forms different organic products, depending upon the reaction conditions used.

Use different reaction conditions to control the type of reaction.

Commonly tested in exams.

KEY POINT

aqueous conditions	⟶	**alcohol**:	OH^- acts as a **nucleophile**
anhydrous conditions	⟶	**alkene**:	OH^- acts as a **base**

The ability to control *which* reaction takes place solely by changing the conditions is an important tool for the synthetic chemist.

Progress check

1 What is meant by the term nucleophilic substitution?

2 Why does 1-bromopropane react with nucleophiles but propane does not?

3 Why do bromoalkanes react more readily than chloroalkanes?

4 What are the organic products for the reactions of 1-bromopropane with:
(a) $NaOH/H_2O$; (b) NH_3/ethanol; (c) NaOH/ethanol?

1 Replacement of an atom or group in an organic molecule by a nucleophile.
2 C–Br bond is polar, C–H bonds are not polar.
3 C–Br bond is weaker than C–Cl bond.
4 (a) $CH_3CH_2CH_2OH$ (b) $CH_3CH_2CH_2NH_2$ (c) $CH_3CH{=}CH_2$.

4.7 Analysis

After studying this section you should be able to:

- *understand that infra-red spectroscopy can be used to identify functional groups*
- *interpret a simple infra-red spectrum*
- *use a mass spectrum to determine the relative molecular mass of an organic molecule*

LEARNING SUMMARY

Using infra-red (IR) spectroscopy in analysis

AQA M2

AQA (A2) M4

Basic principles

Bonds in molecules naturally vibrate. Some bonds in molecules increase their vibrations by absorbing energy from IR radiation. Different bonds absorb different frequencies of IR radiation.

The frequency of IR absorption is measured in wavenumbers, units: cm^{-1}.

An IR spectrum is obtained by passing a range of IR frequencies through a compound. As energy is taken in, **absorption peaks** are produced. The frequencies of the absorption peaks can be matched to those of known bonds to identify structural features in an unknown compound.

IR radiation has less energy than visible light.

KEY POINT: IR spectroscopy is useful for identifying the functional groups in a molecule.

Important IR absorptions

You don't need to learn the absorption frequencies – the data is provided.

bond	*functional group*	*wavenumber/cm^{-1}*
O–H	hydrogen bonded in alcohols	3230 – 3550
N–H	amines	3100 – 3500
C–H	organic compound with a C–H bond	2850 – 3100
O–H	hydrogen bonded in carboxylic acids	2500 – 3300 (broad)
C≡N	nitriles	2200 – 2260
C=O	aldehydes, ketones, carboxylic acids, esters	1680 – 1750
C–O	alcohols, esters	1000 – 1300

IR spectroscopy is used in some modern breathalysers for measuring the concentration of blood alcohol. A particular IR absorption identifies the presence of ethanol in the breath and the intensity of the peak is directly related to the ethanol level.

An infra-red spectrum is particularly useful for identifying:

- an **alcohol** from absorption of the O–H bond
- a **carbonyl** compound from absorption of the C=O bond
- a **carboxylic acid** from absorption of the C=O bond **and** broad absorption of the O–H bond.

Interpreting infra-red spectra

AQA M2

AQA (A2) M4

Carbonyl compounds (aldehydes and ketones)

Butanone,

$CH_3COCH_2CH_3$

- C=O absorption 1680 to 1750 cm^{-1}

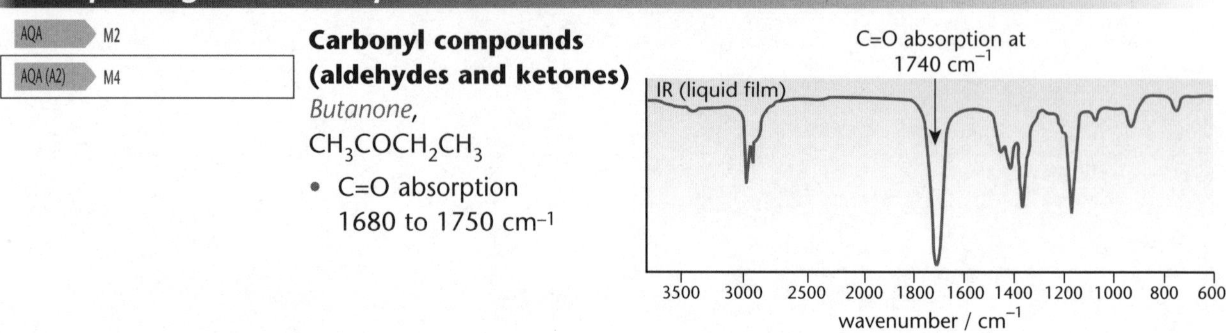

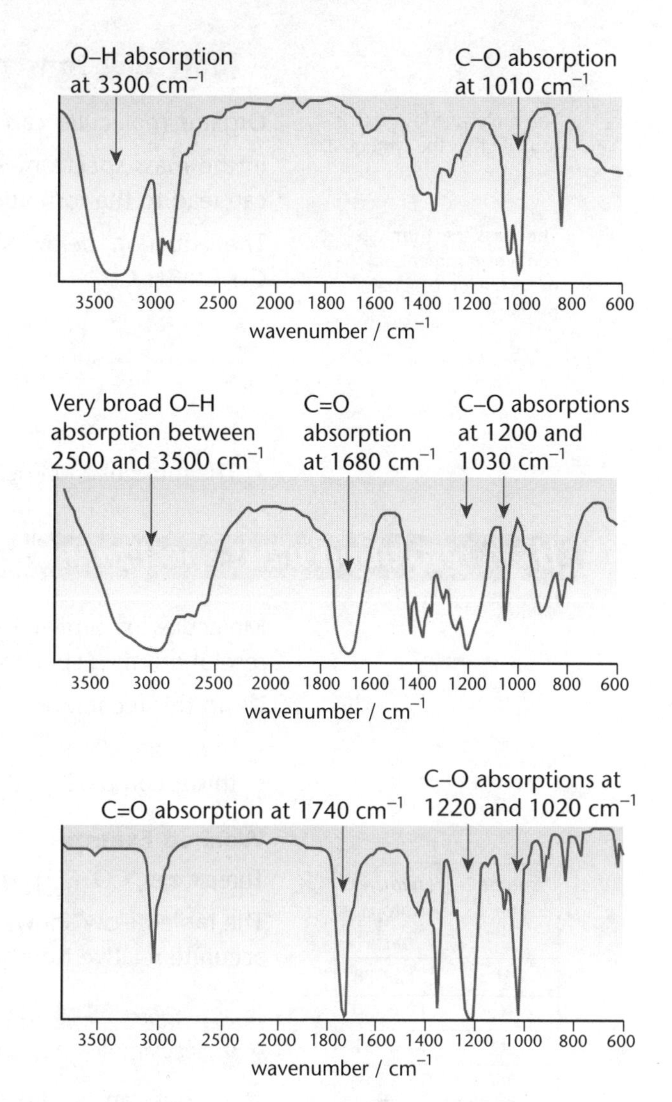

IR spectroscopy is most useful for identifying C=O and O–H bonds.

Look for the distinctive patterns.

Alcohols

Ethanol,
C_2H_5OH

- O–H absorption 3230 to 3500 cm^{-1}
- C–O absorption 1000 to 1300 cm^{-1}

Note that all these molecules contain C–H bonds, which absorb in the range 2840 to 3045 cm^{-1}.

Carboxylic acids

Propanoic acid,
C_2H_5COOH

- Very broad O–H absorption 2500 to 3500 cm^{-1}
- C–O absorption 1680 to 1750 cm^{-1}

There are other organic groups (e.g. N–H, C=C) that absorb IR radiation but the principle of linking the group to the absorption wavenumber is the same.

Ester

Ethyl ethanoate,
$CH_3COOC_2H_5$

- C=O absorption 1680 to 1750 cm^{-1}
- C–O absorption 1000 to 1300 cm^{-1}

The fingerprint region is unique for a particular compound.

Fingerprint region

- Between 1000 and 1550 cm^{-1}

Many spectra show a complex pattern of absorption in this range.

- This pattern can allow the compound to be identified by comparing its spectrum with spectra of known compounds.

Progress check

1 Ethanol, CH_3CH_2OH was oxidised to ethanal, CH_3CHO and then to ethanoic acid, CH_3COOH. How could you use IR spectroscopy to follow the course of this reaction?

1 Ethanol absorbs at about 3300 cm^{-1} (OH); ethanal absorbs at about 1700 cm^{-1} (C=O); ethanoic acid absorbs at about 1700 cm^{-1} (C=O) and 2500–3500 cm^{-1} (broad) (OH from COOH group).

Mass spectrometry

AQA M2

AQA (A2) M4

Mass spectrometry can be used to determine relative atomic masses from a mass spectrum (see page 23). Mass spectrometry is also used to determine relative molecular masses and to identify the molecular structures of organic compounds.

Molecular ions

The molecular ion peak is usually given the symbol M.

The mass spectrum also contains fragments of the molecule ion (see detail below).

Organic molecules can be analysed using mass spectrometry.

In the mass spectrometer, organic molecules are bombarded with electrons. This can lead to the formation a **molecular ion**.

The equation below shows the formation of a molecular ion from butanone, $CH_3COCH_2CH_3$.

$$H_3C{-}C({=}O){-}CH_2CH_3 + e^- \longrightarrow [H_3C{-}C({=}O){-}CH_2CH_3]^+ + 2e^-$$

molecular ion, *m/z*: 72

- The molecular ion peak, M, provides the relative molecular mass of the compound.

High resolution mass spectrometry

AQA M2

Molecules of similar relative molecular masses can be distinguished using high resolution mass spectrometry.

Using this technique:

- the molecular mass is obtained with a very accurate value
- this is compared with very accurate relative isotopic masses.

isotope	*relative isotopic mass*
^{1}H	1.0078
^{12}C	12.0000
^{14}N	14.0031
^{16}O	15.9949

Worked Example

Three gases, CO, C_2H_4 and N_2, each have an approximate relative molecular mass of 28.

The table below shows accurate molecular masses for these gases, calculated from accurate relative isotopic masses (see the margin).

gas	*CO*	*C_2H_4*	*N_2*
molecular mass	27.9949	28.0312	28.0062

The gases can be distinguished by matching the *m/z* values from high resolution mass spectrometry with the calculated values.

4.8 Chemistry in the environment

After studying this section you should be able to:

- *outline the causes of global warming and strategies to curtail the production of greenhouse gases*
- *explain the depletion of the ozone layer by radicals*
- *outline the main principles of green chemistry*

LEARNING SUMMARY

Global warming

AQA M2

The greenhouse effect

The **greenhouse effect** is a natural process, keeping the Earth at a temperature capable of supporting life. Most IR radiation emitted by the Earth goes back into space. However, greenhouse gases in the atmosphere absorb some of the IR radiation, which is then re-emitted as heat, with some passing back towards the Earth.

- The Earth absorbs energy at the same rate as it radiates energy.
- The equilibrium maintains a steady temperature.

The absorption-emission process keeps the heat close to the surface of the Earth.

The main gases in the air are:
nitrogen 78%
oxygen 21%
argon 0.9%
carbon dioxide 0.04%

Although O_2 and N_2 are the main gases in the atmosphere, their molecules do not change polarity when they vibrate and so they do not absorb IR radiation.

Greenhouse gases

Absorption of IR radiation by **greenhouse gases** causes bonds to vibrate. A molecule only absorbs IR radiation if its polarity changes as the bonds vibrate.

The main absorbers of IR radiation are the C=O, O–H and C–H bonds in H_2O, CO_2 and CH_4. Other IR absorbers include the nitrogen–oxygen bonds in nitrogen oxides such as NO and NO_2 (often referred to as NO_x).

The greenhouse effect of a gas depends on:

- its atmospheric concentration
- its ability to absorb IR radiation.

Climate change

Climate change refers to variations in the Earth's climate over time.

- Over hundreds of thousands of years, natural climate changes have been responsible for ice-ages, followed by warming.
- Over shorter periods of time, natural climate change can be the result of factors such as sunspot activity and the presence of volcanic dust in the atmosphere.

The Earth has always experienced **natural climate** change.

Scientific evidence has shown that substantial climate change is now taking place and that the Earth's climate is warming. Scientists agree that the production of greenhouse gases from **human activities** is the primary factor driving **climate change**.

- Human activity is producing **more** greenhouse gases, particularly CO_2.
- Increased proportions of greenhouse gases are upsetting the sensitive, natural balance of nature, resulting in man-made global warming.

Climate change derived from human activities is known as **anthropogenic** climate change, as opposed to natural climate change without human influences.

Computer modelling has predicted significant problems, e.g. melting ice caps, rising sea levels and more extreme storms and flooding.

International treaties, such as the Kyoto Protocol, have set targets for reductions in CO_2 production.

Carbon neutrality

AQA M2

The proportion of CO_2 in the atmosphere has increased more in the last hundred years than at any other time in the Earth's history. This increase is blamed on human activities. Carbon neutrality is seen as an important step in curbing global warming by re-balancing the CO_2 in the atmosphere.

> Carbon neutrality can be achieved in various ways, for example:
>
> changing from finite fuels, such as oil, to renewable fuels based on plants that can be grown annually, such as ethanol and rape-seed oil
>
> cutting down fewer trees and growing more trees to take in CO_2
>
> using less fuel and more public transport
>
> carbon storage and capture (CSS)
>
> improving the efficiency of engines.

KEY POINT

Carbon neutral refers to 'an activity that has no net annual carbon (greenhouse gas) emissions to the atmosphere'.

For carbon neutrality, amounts of CO_2 released into the atmosphere by human activities, such as from burning fuels, is balanced by the **same amounts** of CO_2 being taken in **from** the atmosphere during plant growth.

Alternatively, CO_2 produced during combustion could be stored or 'sequestered' by **carbon storage and capture** (CSS). This can achieved by:

- removal of waste carbon dioxide as a liquid, injected deep in the oceans
- storage as a gas in deep geological formations
- reacting carbon dioxide with metal oxides to form stable carbonate minerals.

This *trapping* of CO_2 emissions would substantially reduce CO_2 emissions from power stations. CSS could make the combustion of fossil fuels carbon neutral.

For transport, carbon neutrality can be approached by use of **biofuels**.

> Although biofuels are often assumed to be carbon neutral, this is not strictly true – additional energy is required for transportation of the fuels, production of fertilisers, etc.

KEY POINT

A **biofuel** is a liquid or gas transportation fuel derived from biomass.

- Biofuels are **renewable**. Once used, fresh fuel can be made again from plants, grown in a short length of time.
- **Finite** resources, such as oil and coal take millions of years to form. Once used, fossil fuels are spent – they are non-renewable.

The best sources of biofuels are alcohols, such as ethanol and methanol, and vegetable oils, such as the rape-seed oil. Biofuels can be used with petrol to fuel cars.

> **Carbon footprint**
>
> As with carbon neutrality, 'carbon footprint' is used to quantify CO_2 emissions.
>
> Carbon footprint is the total amount of CO_2 and other greenhouse gases emitted over the full life cycle of a product or service.

The carbon neutrality of bio-ethanol is achieved in the following way:

- Growing plants produces sugars, taking in CO_2 from the atmosphere.
- The sugars are fermented with yeast, making bio-ethanol.
- The bio-ethanol is burned as a fuel, releasing CO_2 into the atmosphere.
- More sugars are made to form fresh bio-ethanol.

Alternative fuels, such as hydrogen, would produce no CO_2 emissions but the processes required to produce H_2 as a fuel are currently energy intensive. Hydrogen fuels remain a goal for the future, particularly as a fuel for cars.

Depletion of ozone

AQA M2

Depletion of ozone is caused by **radicals**, particularly Cl and NO.

Cl and NO radicals have been formed from human activity, affecting the natural balance that maintains the ozone layer.

> The ozone layer is beneficial for life on Earth.

Cl radicals are formed by the action of **UV radiation** on **CFCs**, which were once used in aerosols and as refrigerants.

e.g. $CClF_3 \longrightarrow Cl + CF_3$

Oxides of nitrogen (NO_x) are formed from thunderstorms and aircraft flying through the upper atmosphere.

e.g. $N_2 + O_2 \longrightarrow 2NO$

The Cl and NO radicals are recycled in chain reactions, which continue many thousand of times until two radicals collide, terminating the reaction. A single Cl radical can remove thousands of O_3 molecules from the ozone layer.

Both processes lead to depletion of ozone in the stratosphere:

stage 1 $O_3 + Cl\bullet \longrightarrow ClO\bullet + O_2$

stage 2 $ClO\bullet + O \longrightarrow O_2 + Cl\bullet$

overall: $O_3 + O \longrightarrow 2O_2$

stage 1 $O_3 + NO\bullet \longrightarrow NO_2\bullet + O_2$

stage 2 $NO_2\bullet + O \longrightarrow O_2 + NO\bullet$

overall: $O_3 + O \longrightarrow 2O_2$

In the processes above, Cl and NO act as catalysts.

- They are used up in the first stage.
- They are regenerated in the second stage.

Overall, Cl and NO radicals are **not consumed**.

Benefits of CFC use versus damage to ozone

In the 1930s, CFCs were seen as being the ideal chemicals for refrigerants, aerosols and as blowing agents for 'expanded poly(styrene)'.

CFCs have the ideal properties for these uses: high volatility, non-toxicity, non-flammability, no smell and extreme unreactivity.

This unreactivity causes the problem.

- Most pollutants are broken down naturally in the lower atmosphere. However, unreactive CFCs diffuse over many years up to the upper atmosphere.
- High-energy UV radiation in the stratosphere finally breaks down the CFCs, forming Cl radicals. The highly reactive Cl radicals attack the ozone.

In the 1970s and 1980s, scientists discovered that Cl radicals were responsible for thinning of the ozone layer.

- Most governments have signed up to the Montreal Protocol, which sets targets for reduced use of CFCs.
- Since the early 2000s, the use of CFCs has been banned internationally. Although replacement chemicals are now in use, many old refrigerators contain CFCs and these have to be disposed of carefully by collecting the CFCs.

It may take a century before the ozone layer fully recovers.

Progress check

1 State three greenhouse gases and the bonds responsible for IR absorption.

2 Outline how the use of bio-ethanol approaches carbon neutrality.

3 Write equations to show how the presence of CFC in the stratosphere depletes the ozone layer.

1 H_2O (O–H); CO_2 (C=O); CH_4 (C–H).

2 Growing plants produces sugars, taking in CO_2 from the atmosphere.
The sugars are fermented with yeast, making bio-ethanol.
The bio-ethanol is burned as a fuel, releasing CO_2 into the atmosphere
Overall, the CO_2 taken in during growth = CO_2 released to the atmosphere (although there will be some extra CO_2 released from transportation and fertiliser production.)

3 $CClF_3 \longrightarrow Cl + CF_3$

stage 1 $O_3 + Cl\bullet \longrightarrow ClO\bullet + O_2$

stage 2 $ClO\bullet + O \longrightarrow O_2 + Cl\bullet$

overall: $O_3 + O \longrightarrow 2O_2$

Sample question and model answer

For open-ended questions of this type, there are usually more marking points than the total – you have some breathing space.

1

(a) Explain, using examples, the following terms used in organic chemistry.

(i) *general formula*

This is the simplest algebraic representation for any member of a series of organic compounds, e.g. *alkanes*: C_nH_{2n+2} ✓

(ii) *homologous series*

This is a series containing compounds with similar chemical properties ✓ with each successive member differing by CH_2 ✓, e.g. 'alkanes' ✓

Do make sure that you choose valid examples. In the actual examination, many candidates managed to confuse homolytic and heterolytic fission – poor preparation!

(iii) *homolytic fission* and *heterolytic fission*

Fission means the breaking of a bond. ✓

In homolytic fission, one electron from the bond is released to each atom with radicals forming. ✓ e.g. $Cl_2 \longrightarrow 2Cl\bullet$ ✓

In heterolytic fission, both electrons in the bonding pair are released to one of the atoms with ions forming ✓ e.g. $Br_2 \longrightarrow Br^+ + Br^-$ ✓

9 marks ⟶ [7]

Always explain your working. This type of problem is often easier than it looks.

You should look back at the question many times to check that you have answered all parts.

(b) Cracking of the unbranched compound **A**, C_6H_{14}, produced the saturated compound **B** and the unsaturated hydrocarbon **C** (M_r, 42). Compound **B** reacted with bromine in UV light to form a monobrominated compound **D** and an acidic gas **E**. Compound **C** reacted with hydrogen bromide to form a mixture of two compounds **F** and **G**.

(i) Use this evidence to suggest the identity of each of compounds **A** to **G**. Include equations for the reactions in your answer.

A is unbranched. ∴ **A** = $CH_3CH_2CH_2CH_2CH_2CH_3$ ✓

C has M_r = 42; must have a double bond: must contain 3 carbons ∴ **C** = C_3H_6. ✓

Notice that everything 'works': questions are designed in this way. If you find yourself with 'strange' structures containing functional groups that you haven't seen before, you have probably made a mistake. This is the time to check back through your logic.

Cracking of A produces an alkane and an alkene, **B** is an alkane: **B** = C_3H_8 ✓

The equation is: $C_6H_{14} \longrightarrow C_3H_6 + C_3H_8$ ✓

Alkanes react with bromine in UV by substitution. ∴ **D** = C_3H_7Br ✓, **E** = HBr ✓

The equation is: $C_3H_8 + Br_2 \longrightarrow C_3H_7Br + HBr$ ✓

Alkenes react with bromine by addition. HBr can be added across a double bond in two ways to give: **F** and **G** as $CH_3CH_2CH_2Br$ ✓ and $CH_3CHBrCH_3$. ✓

The equation is: $C_3H_6 + HBr \longrightarrow C_3H_7Br$ ✓

(ii) Predict the structure of the polymer that could be formed from compound **C**.

When drawing a polymer: 'side bonds' are always needed. Always show the repeat unit.

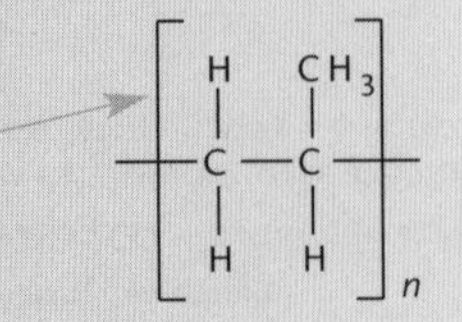

✓✓(1 for no double bond, 1 for correct skeleton) [14]

Practice examination questions

1 There are four structural isomers with the molecular formula $C_3H_6Cl_2$.

(a) (i) Explain the term *structural isomers*.

(ii) Show the structural formulae of the structural isomers of $C_3H_6Cl_2$.

[5]

(b) Some propane was reacted with chlorine to form a mixture of isomers.

(i) What conditions are required for this reaction to take place?

(ii) Two structural isomers, **A** and **B**, were separated from the mixture. These isomers had a molar mass of 78.5 g mol^{-1}. Deduce the molecular formula of these two isomers.

(iii) Draw the structural formulae of **A** and **B** and name each compound.

[6]
[Total: 11]

2 1,2-Dichloroethene has *E* and *Z* isomers.

(a) (i) Explain the two conditions required for an organic molecule to have *E* and *Z* isomers.

(ii) What is the difference between *E* and *Z* isomers?

(iii) Draw and label the *E* and *Z* isomers of 1,2-dichloroethene.

[5]

(b) The *E* and *Z* isomers of 1,2-dichloroethene can both be reduced to 1,2-dichloroethane. What are the reagents and conditions required?

[2]

(c) The *E/Z* isomers of $C_2H_2Cl_2$ polymerise to form poly(1,2-dichloroethene). Draw a short section of poly(1,2-dichloroethene), showing **two** repeat units.

[2]
[Total: 9]

3 The flowchart below shows a reaction sequence starting from but-2-ene.

$$CH_3CH{=}CHCH_3 \xrightarrow{HBr} \mathbf{A} \xrightarrow[\text{heat}]{OH^-(aq)} \mathbf{B} \xrightarrow[\text{heat}]{Cr_2O_7^{2-}/H^+} \mathbf{C}$$

(a) Draw a short section of the addition polymer formed by but-2-ene showing two repeat units.

[2]

(b) Identify compounds **A–C** in the flowchart above.

[3]

(c) Name the types of reaction taking place in the flowchart above and state the role of the reagents.

(i) $CH_3CH{=}CHCH_3 \longrightarrow$ **A**

(ii) **A** $\longrightarrow$ **B**

(iii) **B** $\longrightarrow$ **C**

[6]

(d) What reagents and conditions would be needed to convert but-2-ene directly into compound **B**?

[2]
[Total: 13]

Practice examination questions

4 A branched alkene **A** was reduced with hydrogen to form a compound **B**. Compound **B** had the composition by mass of C, 82.8 %; H, 17.2 % (M_r = 58). Compound **A** was reacted with steam with an acid catalyst to produce a mixture of two structural isomers **C** and **D**.

(a) Calculate the empirical and molecular formula of compound **B**. [3]

(b) Show structural formulae for compounds **A–D**. [4]

(c) Write an equation for each reaction. [2]

[Total: 9]

5 (a) There are four structural isomers of molecular formula C_4H_9Br. The structural formulae of two of these isomers are:

- Compound **A**: $CH_3CH_2CH_2CH_2Br$
- Compound **B**: $(CH_3)_2CHCH_2Br$.

(i) Show the remaining two structural isomers.

(ii) Give the name of **B**.

[3]

(b) All four structural isomers of C_4H_9Br react with hot aqueous sodium hydroxide.

(i) What type of reaction is taking place?

(ii) Draw the structural formula of the organic product formed by the reaction of **B** with hot aqueous sodium hydroxide.

[2]

(c) Ethene, C_2H_4, reacts with bromine.

(i) State the name of the mechanism involved.

(ii) Show the mechanism for this reaction.

[4]

[Total: 9]

Practice examination answers

Chapter 1 Atoms, moles and reactions

1 (a)

particle	*relative mass*	*relative charge*	
proton	1	1+	✓
neutron	1	0	✓
electron	1/2000	1–	✓

[3]

(b) (i) neutron ✓; ^{13}C has 1 more neutron ✓

(ii) electron ✓; $^{16}O^{2-}$ has 2 more electrons ✓

(iii) proton ✓; $^{24}Mg^{2+}$ has 1 more proton ✓ [6]

[Total: 9]

2 (a) Difference: different numbers of neutrons. ✓

Similarities: same numbers of protons ✓ and electrons. ✓ [3]

(b) ^{11}B ✓ [1]

(c) (i) $1s^22s^22p^63s^23p^6$ ✓; (ii) p block ✓ [2]

(d) $Si(g) \longrightarrow Si^{+}(g) + e^{-}$; ✓✓. [6]

[Total: 12]

3 (a) (i) 0.0170 ✓; (ii) 0.0850 mol dm^{-3} ✓✓;

(iii) 204 cm^3 using 24 dm^3; 211 cm^3 using $pV=nRT$ (AQA) ✓✓✓ [6]

(b) 0.0741 mol dm^{-3} ✓✓✓ [3]

[Total: 9]

4 (a) (i) 0.0145 mol ✓; (ii) 0.0290 mol ✓; (iii) 23.2 cm^3 ✓ [3]

(b) $M(Na_2CO_3)$ = 106 g mol^{-1} ✓

mass Na_2CO_3 = 0.0145 x 106 = 1.537 g ✓ [2]

(c) Mg is oxidised from 0 to +2 ✓; H is reduced from +1 to 0 ✓ [2]

[Total: 7]

5 (a) (i) 0.0625 ✓; (ii) $M(NaN_3)$ = 65 g mol^{-1} ✓;

moles NaN_3 = 0.0417 mol ✓; mass NaN_3 = 0.0417 x 65 = 2.71 g ✓ [4]

(b) (i) concentration = $0.0417 \times \frac{1000}{50.0}$ ✓ = 0.834 mol dm^{-3} ✓

(ii) Na is oxidised from 0 to +1 ✓; H is reduced from +1 to 0 ✓ [4]

[Total: 8]

Chapter 2 Bonding, structure and the Periodic Table

1 (a) A covalent bond is a shared ✓ pair ✓ of electrons. [2]

(b) In a dative covalent bond, the same atom provides both electrons for the bonded pair ✓. The O of a water molecule uses a lone pair of electrons to form a dative covalent bond with H^+ to form an H_3O^+ ion ✓. [2]

(c) H_2O: 104.5° ✓; H_3O^+: 107° ✓. In H_3O^+, there is repulsion between three bonded pairs of electrons and one lone pair of electrons ✓. The lone pair repels more than the bonded pairs, reducing the tetrahedral bond angle by 2.5° ✓. [4]

(d) Hydrogen bonding ✓. This arises from dipole-dipole attraction ✓ between $H^{\delta+}$ of a water molecule and $O^{\delta-}$ on a different water molecule ✓. [3]

[Total: 11]

2 Electron pairs repel one another ✓ and assume a position as **far away** from one another as possible ✓. Lone pairs repel more than bonded pairs ✓. Diagrams are shown below:

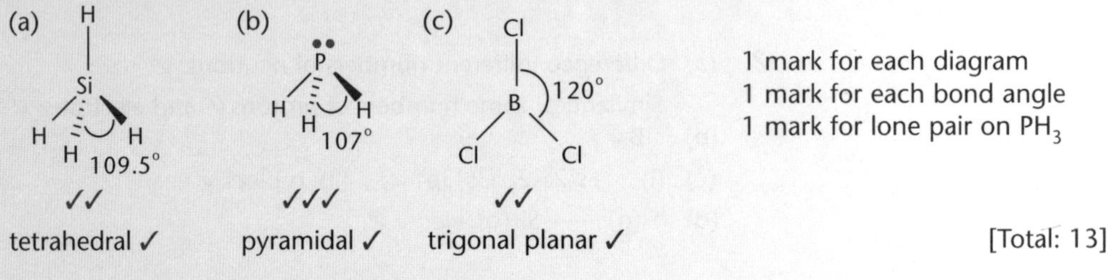

tetrahedral ✓ pyramidal ✓ trigonal planar ✓ [Total: 13]

3 (a) (i) covalent ✓; (ii) van der Waals' forces ✓. [2]

(b) Weak van der Waals' (or induced dipole-dipole) forces break on vaporising ✓. [1]

(c) Strong ✓ ionic bonds ✓ hold the lattice together. [2]

(d) Ions vibrate faster ✓. [1]

(e) On melting, the ionic bonds remain close together and are just loosened from their rigid lattice positions ✓. On vaporising, the ions must be separated, breaking the ionic attraction ✓. [2]

[Total: 8]

4 (a) Water has hydrogen bonding ✓. This is dipole-dipole attraction ✓ between $H^{\delta+}$ of a water molecule and a lone pair ✓ on $O^{\delta-}$ on a different water molecule ✓. [4]

(b) HCl has dipole-dipole interactions ✓ between $H^{\delta+}$ of a HCl molecule ✓ and $Cl^{\delta-}$ on a different HCl molecule ✓. any 2 ⟶ [2]

(c) Kr has van der Waals' forces only ✓. These are caused by oscillating dipoles ✓ which cause induced dipoles on another Kr atom ✓.
H-bonding > dipole-dipole > van der Waals' ✓. The intermolecular forces are broken on boiling ✓. The stronger the intermolecular forces, the higher the boiling point ✓. any 4 ⟶ [4]

[Total:10]

5 Across Period 3, the atomic radii decrease ✓ because there are more protons in the nucleus across the period ✓ increasing the attraction ✓ on the electrons in the same outer shell ✓.

Down Group 2, the atomic radii increase ✓ as more shells are being added ✓. The shielding effect increases as the number of inner shells increases ✓. The attraction experienced by the outer electrons from the nucleus decreases ✓. [Total: 8]

6 (a) (i) $Ba(s) + 2H_2O(l) \longrightarrow Ba(OH)_2(aq) + H_2(g)$ ✓

(ii) $2Mg(s) + O_2(g) \longrightarrow 2MgO$ ✓

(iii) $Mg(s) + 2HCl(aq) \longrightarrow MgCl_2(aq) + H_2(g)$ ✓ [3]

(b) (i) On descending, outer electrons are further from the nucleus ✓ with greater shielding ✓. Therefore less attraction on outer electrons which are lost more easily ✓.

(ii) Reaction with $H_2O/O_2/Cl_2$/named acid ✓. [4]

(c) (i) Very high melting point/strong forces holding lattice together ✓.

(ii) CaO: cement ✓; $Mg(OH)_2$: indigestion remedy ✓ *many others possible.* [3]

[Total: 10]

7 (a) On descending Group 7, the atomic radii increase ✓ resulting in less nuclear attraction at the edge of the atom ✓. There are more electron shells between the nucleus and the edge of the atom to shield the nuclear charge ✓. The nuclear attraction at the edge of the atom decreases down the group ✓. ∴ the oxidising power of the Group 7 elements decreases down the group ✓. [5]

(b) (i) $Cl_2 \longrightarrow NaCl$ (ox no: Cl 0 → −1) ✓ reduction ✓;
$Cl_2 \longrightarrow NaClO_3$ (ox no: Cl 0 → +5) ✓ oxidation ✓

(ii) 2.7 mol Cl_2 reacts ✓; 0.90 mol $NaClO_3$ forms ✓; 9.6 g ✓

(iii) $Cl_2 + 2NaOH \longrightarrow NaOCl + NaCl + H_2O$ ✓; bleach ✓ [9]

[Total: 14]

8 (a) (i) Electronegativity is a measure of the attraction of an atom in a molecule for the pair of electrons in a covalent bond ✓.

(ii) Electronegativity decreases from fluorine to iodine ✓. The atomic radii increase down the group ✓. The shielding effect also increases down the group as the number of electron shells increases ✓. The smaller the halogen atom, the greater the nuclear attraction experienced by the bonding electrons and the greater the electronegativity ✓. [5]

(b) The boiling points increase from fluorine to iodine ✓. The number of electrons increases ✓, increasing the van der Waals' forces between molecules ✓. [3]

(c) With aqueous potassium chloride, there is no change ✓. With aqueous potassium iodide, the solution turns dark brown ✓ as iodine is displaced ✓.
$Br_2(aq) + 2I^-(aq) \longrightarrow 2Br^-(aq) + I_2(aq)$ ✓ [4]

(d) Bromine is a weaker oxidising agent than chlorine because it does not oxidise chloride ions to chlorine ✓. Bromine is a stronger oxidising agent than iodine because it oxidises iodide ions to iodine ✓. [2]

[Total: 14]

Chapter 3 Energetics, rates and equilibrium

1 (a) $2Na(s) + C(s) + 1\frac{1}{2}O_2(g) \longrightarrow Na_2CO_3(s)$ ✓✓ [1, equation; 1, state symbols] [2]

(b) 100 kPa and 298 K ✓ [1]

(c) +90 kJ mol^{-1} ✓✓✓ [3]

[Total: 6]

2 (a) (i) The enthalpy change that takes place when one mole of a substance ✓ reacts completely with oxygen ✓ under standard conditions, all reactants and products being in their standard states ✓.

(ii) 298 K ✓ [4]

(b) (i) $CH_3CH_2CH_2OH(l) + 4\frac{1}{2}O_2(g) \longrightarrow 3CO_2(g) + 4H_2O(l)$ ✓

(ii) −2023 kJ mol^{-1} ✓✓✓ [4]

[Total: 8]

3

(a) The enthalpy change that takes place when one mole of a compound ✓ in its standard states is formed from its constituent elements ✓ in their standard states under standard conditions ✓. [3]

(b) $C(s) + 2H_2(g) \longrightarrow CH_4(g)$ ✓✓ [1, equation; 1, state symbols] [2]

(c) −76 kJ mol^{-1} ✓✓✓ [3]

(d) C–H: +412 kJ mol^{-1} ✓; C–C: +348 kJ mol^{-1} ✓✓ [3]

[Total: 11]

4 (a) (i) ✓✓✓; (ii) ✓✓

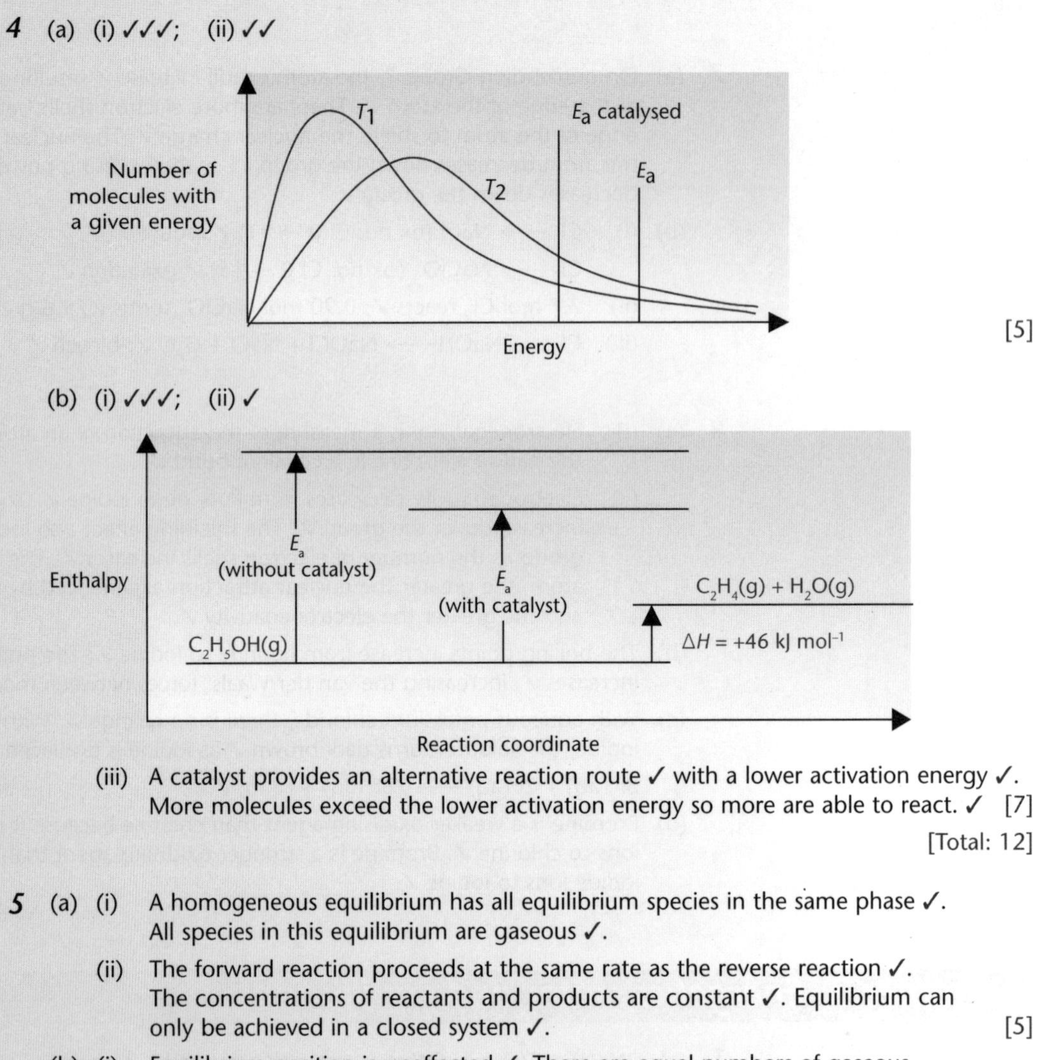

[5]

(b) (i) ✓✓✓; (ii) ✓

(iii) A catalyst provides an alternative reaction route ✓ with a lower activation energy ✓. More molecules exceed the lower activation energy so more are able to react. ✓ [7]

[Total: 12]

5 (a) (i) A homogeneous equilibrium has all equilibrium species in the same phase ✓. All species in this equilibrium are gaseous ✓.

(ii) The forward reaction proceeds at the same rate as the reverse reaction ✓. The concentrations of reactants and products are constant ✓. Equilibrium can only be achieved in a closed system ✓. [5]

(b) (i) Equilibrium position is unaffected ✓. There are equal numbers of gaseous moles on either side of the equilibrium ✓.

(ii) Rate increases ✓. Increasing the pressure also increases the concentration ✓. [4]

(c) Endothermic ✓. With increased temperature, an equilibrium shifts in the endothermic direction ✓. [2]

[Total: 11]

Chapter 4 Organic chemistry, analysis and the environment

1 (a) (i) Same **molecular** formula with a different arrangement/structural formula ✓.

(ii) $ClCH_2CH_2CH_2Cl$, $CH_3CHClCH_2Cl$, $CH_3CH_2CHCl_2$, $CH_3CCl_2CH_3$ ✓✓✓✓ [5]

(b) (i) UV / sunlight ✓; (ii) C_3H_7Cl ✓

(iii) $CH_3CH_2CH_2Cl$ ✓ 1-chloropropane ✓
$CH_3CHClCH_3$ ✓ 2-chloropropane ✓

[6]

[Total: 11]

Organic chemistry, analysis and the environment

2 (a) (i) There must be a C=C double bond that prevents rotation. ✓
There must be different groups attached to each carbon atom in the C=C double bond. ✓
(ii) *E*: groups are on opposite sides of double bond.
Z: groups are on same side of double bond. ✓
(iii) correct structures ✓✓

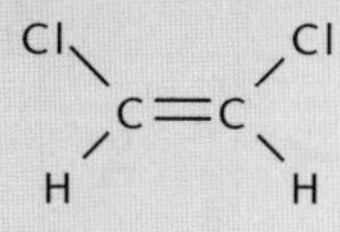

[5]

(b) H_2 ✓ and a named catalyst/Ni/Pt/Pd ✓. [2]

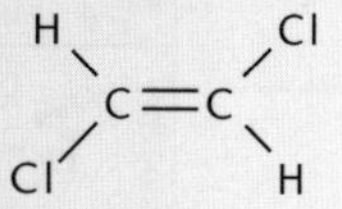

(c)

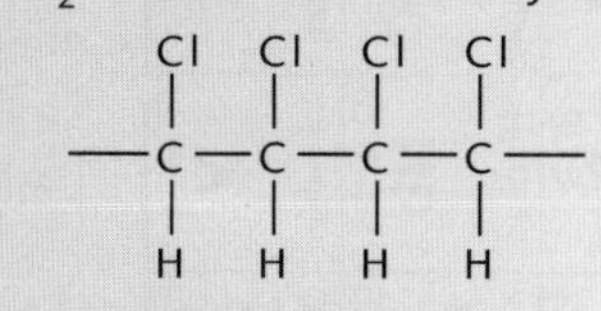

2 repeat units ✓ with side links ✓

[2]
[Total: 9]

3 (a)

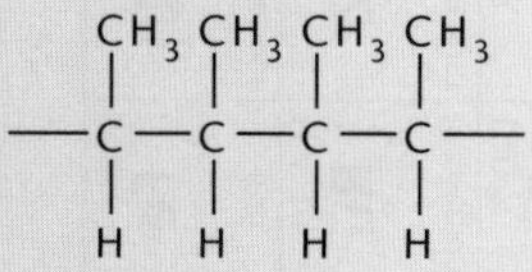

```
  CH3 CH3 CH3 CH3
   |   |   |   |
 —C —C —C —C—
   |   |   |   |
   H   H   H   H
```

no double bond ✓, correct skeleton ✓ [2]

(b) **A** = $CH_3CH_2CHBrCH_3$ ✓; **B** = $CH_3CH_2CHOHCH_3$ ✓; **C** = $CH_3CH_2COCH_3$ ✓ [3]

(c) (i) addition ✓, electrophile ✓;
(ii) substitution ✓, nucleophile ✓;
(iii) oxidation ✓, oxidising agent ✓. [6]

(d) steam ✓ and H_3PO_4 ✓ [2]

[Total: 13]

4 (a) Empirical formula of **B** = C_2H_5 ✓✓; Molecular formula of **B** = C_4H_{10} ✓ [3]

(b) **A** = $(CH_3)_2C{=}CH_2$ ✓; **B** = $(CH_3)_2CHCH_3$ ✓;
C and **D** = $(CH_3)_2CHCH_2OH$ ✓ and $(CH_3)_2COHCH_3$ ✓; [4]

(c) $C_4H_8 + H_2 \longrightarrow C_4H_{10}$ ✓
$C_4H_8 + H_2O \longrightarrow C_4H_9OH$ ✓ [2]

[Total: 9]

5 (a) (i) $CH_3CH_2CHBrCH_3$ ✓; $(CH_3)_3CBr$ ✓.
(ii) 1-bromo-2-methylpropane ✓ [3]

(b) (i) nucleophilic substitution ✓; (ii) $(CH_3)_2CHCH_2OH$ ✓ [2]

(c) (i) electrophilic addition ✓
(ii) arrows stage 1 ✓; arrow stage 2 ✓; carbocation ✓

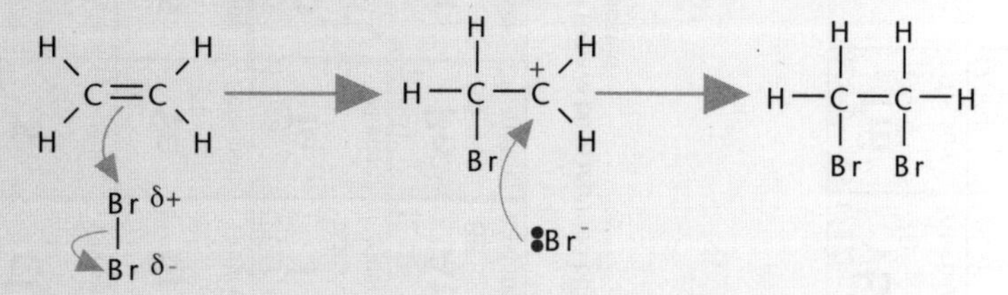

[4]
[Total: 9]

The Periodic Table

Key: 1.0 H 1 Hydrogen — relative atomic mass (1.0); atomic symbol (H); atomic number (1); name (Hydrogen)

1	2											3	4	5	6	7	0
																	4.0 He 2 Helium
6.9 Li 3 Lithium	9.0 Be 4 Beryllium											10.8 B 5 Boron	12.0 C 6 Carbon	14.0 N 7 Nitrogen	16.0 O 8 Oxygen	19.0 F 9 Fluorine	20.2 Ne 10 Neon
23.0 Na 11 Sodium	24.3 Mg 12 Magnesium											27.0 Al 13 Aluminium	28.1 Si 14 Silicon	31.0 P 15 Phosphorus	32.1 S 16 Sulfur	35.5 Cl 17 Chlorine	39.9 Ar 18 Argon
39.1 K 19 Potassium	40.1 Ca 20 Calcium	45.0 Sc 21 Scandium	47.9 Ti 22 Titanium	50.9 V 23 Vanadium	52.0 Cr 24 Chromium	54.9 Mn 25 Manganese	55.8 Fe 26 Iron	58.9 Co 27 Cobalt	58.7 Ni 28 Nickel	63.5 Cu 29 Copper	65.4 Zn 30 Zinc	69.7 Ga 31 Gallium	72.6 Ge 32 Germanium	74.9 As 33 Arsenic	79.0 Se 34 Selenium	79.9 Br 35 Bromine	83.8 Kr 36 Krypton
85.5 Rb 37 Rubidium	87.6 Sr 38 Strontium	88.9 Y 39 Yttrium	91.2 Zr 40 Zirconium	92.9 Nb 41 Niobium	95.9 Mo 42 Molybdenum	[98] Tc 43 Technetium	101.1 Ru 44 Ruthenium	102.9 Rh 45 Rhodium	106.4 Pd 46 Palladium	107.9 Ag 47 Silver	112.4 Cd 48 Cadmium	114.8 In 49 Indium	118.7 Sn 50 Tin	121.8 Sb 51 Antimony	127.6 Te 52 Tellurium	126.9 I 53 Iodine	131.3 Xe 54 Xenon
132.9 Cs 55 Caesium	137.3 Ba 56 Barium	138.9 La* 57 Lanthanum	178.5 Hf 72 Hafnium	180.9 Ta 73 Tantalum	183.8 W 74 Tungsten	186.2 Re 75 Rhenium	190.2 Os 76 Osmium	192.2 Ir 77 Iridium	195.1 Pt 78 Platinum	197.0 Au 79 Gold	200.6 Hg 80 Mercury	204.4 Tl 81 Thallium	207.2 Pb 82 Lead	209.0 Bi 83 Bismuth	[209] Po 84 Polonium	[210] At 85 Astatine	[222] Rn 86 Radon
[223] Fr 87 Francium	[226] Ra 88 Radium	[227] Ac* 89 Actinium	[261] Rf 104 Rutherfordium	[262] Db 105 Dubnium	[266] Sg 106 Seaborgium	[264] Bh 107 Bohrium	[277] Hs 108 Hassium	[268] Mt 109 Meitnerium	[271] Ds 110 Darmstadtium	[272] Rg 111 Roentgenium							

Elements with atomic numbers 112–116 have been reported but not fully authenticated

lanthanides	140.1 Ce 58 Cerium	140.9 Pr 59 Praseodymium	144.2 Nd 60 Neodymium	144.9 Pm 61 Promethium	150.4 Sm 62 Samarium	152.0 Eu 63 Europium	157.2 Gd 64 Gadolinium	158.9 Tb 65 Terbium	162.5 Dy 66 Dysprosium	164.9 Ho 67 Holmium	167.3 Er 68 Erbium	168.9 Tm 69 Thulium	173.0 Yb 70 Ytterbium	175.0 Lu 71 Lutetium
actinides	232.0 Th 90 Thorium	[231] Pa 91 Protactinium	238.1 U 92 Uranium	[237] Np 93 Neptunium	[242] Pu 94 Plutonium	[243] Am 95 Americium	[247] Cm 96 Curium	[245] Bk 97 Berkelium	[251] Cf 98 Californium	[254] Es 99 Einsteinium	[253] Fm 100 Fermium	[256] Md 101 Mendelevium	[254] No 102 Nobelium	[257] Lr 103 Lawrencium

Notes

Notes

Notes

Index